Electro Magnetic Pollution

A Handbook for Mariners

Electro Magnetic Pollution

A Handbook for Mariners

P. Misra

Narosa Publishing House
New Delhi Chennai Mumbai Kolkata

Electro Magnetic Pollution
A Handbook for Mariners
146 pgs. | 60 figs. | 35 tbls.

P. Misra
Mercantile Marine Department
Chennai, India

NAROSA PUBLISHING HOUSE PVT. LTD.

22, Delhi Medical Association Road, Daryaganj, New Delhi 110 002
35-36 Greams Road, Thousand Lights, Chennai 600 006
306 Shiv Centre, Sector 17, Vashi, Navi Mumbai 400 703
2F-2G Shivam Chambers, 53 Syed Amir Ali Avenue, Kolkata 700 019

www.narosa.com

ISBN 978-81-8487-143-2

Published by N.K. Mehra for Narosa Publishing House Pvt Ltd.,
22, Delhi Medical Association Road, Daryaganj, New Delhi 110 002

Printed in India

Dedicated to
All those men and women who endeavored to make EMI, EMC, EMR and EMP and its effects known to the common man.

Preface

Ever since mechanical means of propulsion have been integrated into ships they have started becoming larger, more sophisticated and are able to perform complex tasks. The use of electronic equipment, communication equipment and other sophisticated equipments has become immense. Electronic components in systems are now been used on board for everything from traditional systems to communication and navigation. These electronic components started making a foray into the industry about 6 decades back, and today perform the important role of control, communications and monitoring on board vessels. The emission control requirements of 1970's are recognized as the time when electronics started to make an appearance in a predominant way on ship board systems.

In the last two decades the Electromagnetic Compatibility (EMC) in the design of control, monitoring and communication devices on board ships have become important throughout the world. Many countries today especially industrialized nations do not allow electronic devices to be marketed unless their electromagnetic noise emission have been tested and certified to meet certain limits. As a result, the shipping industries are increasingly showing interest in qualified engineers with EMC knowledge. In the last 10 years the United States Federal Communications Commission (FCC) imposing restrictions of radiated and conducted emissions of digital devices have sensitized users in the shipping industry on the effects and virtues of the EMR and EMC.

Shipping industries throughout the world are keen on EMC compliance because of various safety requirements and because of competitive nature of today's markets. If a certain product does not meet regulatory requirements then it is not accepted for ship building.

There are many areas of interest in EMC which can manifest itself in ship board utility. A ship has very compact spaces for various types of emitters and radiators. Further, a digital product within this environment is sensitive to Electrostatic Discharge (ESD) and when it fails to operate reliably then its marketability is lost. Thus EMC applications on board ships have become a specialized area of engineering and which now is becoming more and more integrated into the ship operations and shipping maintenance cycles.

Unfortunately there are no guidelines in place for continued monitoring of EMC compliance on board Merchant Navy vessels. Other than ensuring the compatibility at the time of fitment, radiation measurements are practically never taken throughout the life of the ship. This is very unlike the Navy where continued monitoring norms are in practice. Consequently, EMC qualified personnel are hard to find, though research bodies and specialized institutions have now come up for taking measurements and providing guidance on EMC compliance. In spite of knowing the hazards of EMR and EMI as well as virtues of EMC, compliance is not yet mandatory. There are no conventions or rules other than national requirements imposed on radiating components on board ships. The awareness of EMR, EMC and its effects within the shipping industries operationally, is also low. Sometimes unexplained medical incidents occur which can be easily proved as a consequence of EMR.

The format of the text is basically structured to sensitize the personnel in the shipping industry of the consequences of EMI and EMR. The basics of its action and how to achieve EMC, which will necessarily become mandatory in a few years, is to ensure safety of personnel and equipment on board.

Chapter – 1 gives an introduction into the basics of EMC, explaining what is EMC and EMR and also reveals sources of such emission. Since most technical personnel need to have a basic understanding of susceptibility and immunity, this chapter deals briefly on the two important issues. The correlation of EMC and development of technology as well as its ingress in shipping systems is explained in very simple terms so that the common user of ship board machinery or somebody who has spent a lot of time in such environment can take sufficient basic precautions.

Chapter 2 further explains the sources and conditions of exposure, the various types of emitters, the role of antennae and types of antennas and the mechanism by which even common wire connections, ladders and stays can be sources of emission. A need was also felt to explain the importance of shields which are normally able to indicate the system or the component to ensure compatibility. Using this, shipping board personnel can device their own shielding methods for rogue emitters.

Chapter 3 deals on the effects of EMI and EMC, the radiation penetration, the effects of strengthening of EMI on board vessels and the effects of EMI and EMR. This chapter devotes some explanations to the effects of EMR on personnel which can be harmful if appropriate precaution is not taken. The consequence of biological hazard effects other than radiation hazard effects on board ships is specialized, interspersed with information from various research findings around the world and the effects it has had on personnel working in closed quarters of such radiators.

Chapter 4 describes in detail the entire process of EMC tests and validation of marine equipments. Since ships have a multiplicity of antenna including natural antennas as well as miles of cables, it is imperative that EMC testing needs to be carried throughout the vessel for compliance. Hence, understanding the radiation characteristics and the principles of compatibility of the instrumentation is essential. This book is a most practical book for the maritime professionals to be thorough with, since most of the principles and practice of EMC testing and analysis is explained here.

Nothing would be complete without attempting simulation and modeling of the compatibility on board ships under construction. Since equipment has to be specialized in very close proximity and cables have to be routed through sensitive areas, the entire process can be simulated and modeled for compatibility which would help the reader to understand the practices that need to be put in place in meeting the requirements. There are many production techniques in place and some of these production techniques have now been used extensively for purpose of forecasting ship safety. A mention is made on the principles of a safety index the process by which the safety normally can be projected, using forecasting methods in which EMC forms an integral part.

Chapter 6 deals mainly with ship board environment and the ship board test and valuation procedures. Since there is extensive cabling on ships, the effects of cabling are discussed in Chapter 7. The ways and means to measure the Radio Frequencies and specific shielding required for cables at different locations is also explained. There are specifications explained in cabling and working on board vessels for attenuating broad band and narrow band noise.

Finally there are annexures to these chapters which give a specification on the classification of cables and cabling parts on ship communication, coupling paths on bridge, flow chart for modeling and the EMC checklist. An information sheet showing the information management is also provided. This book would essentially be a book of reference and handbook for maritime professionals who are either engaged on ships or are participating in ship building, ship repair and complex ship operations including cargo operations. In times to come, when these professionals become more and more aware of the essential requirements for personal safety under EMR conditions, the safety perspective of a ship would also be met and along with it safety concerns would also reduce. This is a first book of its kind for maritime professionals and as more and more EMC awareness and regulations become mandatory, maritime professionals will be prepared to meet the compliance regime in an effective way.

P. Misra

Acknowledgement

There have been many pioneering research papers on Electro Magnetic Radiation and commendable work carried out in India and Abroad. However other than a few specifications for EMI, EMC on board ships, there was no book for reference and study especially for seafarers. The initiative for writing this book would not have been possible but for the relentless efforts being made by my teachers in IIT Kharagpur namely Prof. B.N.Dass, Prof. Emeritus and Prof. Sarkar, from SAMEER, Mumbai. If there is one man who I can single out as the greatest source of motivation and support, it has been Prof. Ajay Chakraborty of IIT, Kharagpur, Kalpana Chawla, Space Centre. His brilliance and approach to make shipping a safer place is probably the greatest spirit behind this outcome. Extensive work has been carried out by my mentor and guide Prof. G.S.N. Raju in the Department of Electronics, Andhra University Visakhapatnam. His assistance in procuring in valuable information in Antenna Radiation, Cable Radiation and the intricacies of Electromagnetic Interference and Compatibility is invaluable. But for his constant guidance and encouragement, this book would not have been possible. Our indebtedness to the laboratory and staff of IIT Kharagpur Electronics Department and Kalpana Chawla Centre for Space Research for having provided valuable Measurements and Analysis of Electro Magnetic Emission on Board Ships. The prime Laboratory work and development of software by Prof. Biswas, Prof. Sahoo and Prof. Amithva Bhattacharjee are the principal inputs in to the analysis of EME on Board Ships. This is to also place on record my acknowledge ment and appreciation of the contribution made by Prof. Sundar and Prof. Avudai Nayagam of Mathematics Department of IIT Chennai for having provided support and solutions to many complex problems.

There are many young researchers who have assisted in compiling this book, special mention being made of Prof. S.C.Ramesh, Department of Electrical and Electronics Engineering of PSN College of Engineering and Technology, Tirunelveli. The encouragement and assistance provided by Prof.P.Suyambu Chairman PSN Group of Institutions during the compilation of Data and Literature for this book, is immense

Lastly I would like to thank my Parents and Wife for having given me immense encouragement courage and support to complete this project which in their humble way they felt would ultimately be of immense value to the ordinary citizens who are ignorant of the subtle effects of electromagnetic radiation. My obeisance to them for having stood by me during the difficult process of compiling this book.

P. Misra

List of Figures

List of Tables

Abbreviations

AC - Alternate Current
AM - Amplitude Modulation
BB - Broad Band
CE - Conducted Emissions
CI - Conducted Immunity
CISPR - Special International Committee on Radio Interference
CM - Common Mode
CW - Continuous Wave
DC - Direct Current
DM - Differential Mode
DNV - Det Norske Veritas
EEC - European Energy Community
EFT - Electrical Fast Transient
EM - Electro Magnetic
EMC - Electro Magnetic Compatibility
EME - Electro Magnetic Emission
EMI - Electro Magnetic Interference
EMP - Electro Magnetic Pollution
EMR - Electro Magnetic Radiation
EPIRBS - Emergency Position-Indicating Radio Beacons
ERP - Effective Radiated Power
ESD - Electro Static Discharge
EUT - Equipment Under Test
eV - electron Volt
FCC - Federal Communications Commission
FM - Frequency Modulation
GPS - Global Positioning System
HFSD - High Flashpoint Speed Diesel
HSD - High Speed Diesel
HVAC - High Voltage Alternate Current
IMO - International Maritime Organization
INIRC - International Non-Ionizing Radiation Committee
IRPA - International Radiation Protection Association

ITU	-	International Telecommunication Union
KHz	-	Kilo Hertz
KV	-	Kilo volt
MHz	-	Mega Hertz
ms	-	milli second
LPDA	-	Log-Periodic Dipole Array
NB	-	Narrow Band
NSSA	-	National Space Society of Australia
OD	-	Outside Diameter
PCB	-	Printed Circuit Board
PWM	-	Pulse Width Modulation
RAF	-	Receiving Antenna Factor
RE	-	Radiated Emissions
RF	-	Radio Frequency
RFI	-	Radio Frequency Interference
RI	-	Radiated Immunity
SCR	-	Silicon Controlled Rectifier
SE	-	Shielding Effectiveness
SI	-	Spark Ignited
SNR	-	Signal to Noise Ratio
TAF	-	Transmitting Antenna Factor
THI	-	Temperature Humidity Index
UWB	-	Ultra Wide Band
VHF	-	Very High Frequency
VSWR	-	Voltage standing Wave Ratio
WHO	-	World Health Organization
WW	-	World War

Contents

Chapter 1

What is EMC?

1.1 INTRODUCTION

Phenomenal growth in the utilization of electrical and electronic devices in Shipping defense, industry, commerce, medicine, research etc, in the present decade has created a high level of electromagnetic pollution in the environment, causing a serious problem in the form of electromagnetic interference (EMI). The EMI always creates an unreliable, multifunctional operation in extremely sensitive electronic equipment used in Shipping defense, space exploration and research, communication, medical, commerce etc.

An awareness of EMI is gaining momentum due to the EEC directive and also due to an increase in the EMI problems encountered by engineers working with communication equipment. The awareness has also spread to electrical and electronic system/equipment manufactures.

In the environment the equipment is exposed to electromagnetic interference (EMI) originated by physical phenomena or generated by various equipment. EMI is somewhat arbitrarily defined to cover the frequency spectrum from about 10Hz to 100 GHz. for radiated emissions. A lower frequency limit of 10 KHz is often used, although EMI can exist in many equipment and systems below this frequency. Except for electrostatic discharge (ESD) there rarely exists a pure DC EMI problem.

In the case of shipping, EM environment will be variable from place, ship to ship and between locations. Estimation of EMI environment in any situation is required before adequate protection methods can be selected which will enable equipment to operate without error in all environments. For example, if control valve operation is initiated automatically by a micro computer during cargo discharge, the equipment must be capable of continuous operation in harbour electromagnetic environment.

Depending on the different environments, a wide variety of interference sources can be encountered. Power converters, switch gears, contactors, relays, welders, radio and television transmitters and mobile radios, are among the most conspicuous EMI sources. Transient disturbances, which occur most frequently, usually for short random periods of time and mostly result from interferences caused by lightning, earth-faults or switching of inductive circuits. These disturbances can have a frequency range from 50 Hz up to a few hundred megahertz with time duration including transients ranging from less than 10 nanoseconds to a few seconds.

The evolution and convergence of multiple technologies provided a means to retrofit monitoring equipment on a distributed shipboard system to achieve a reduction in maintenance and operating costs, and mainly effects on humans.

Advances in sensor technology have improved data collection and management strategies. Wireless communication standards and the proliferation of shipboard computers connected through an Ethernet network have enabled an innovative approach to equipment monitoring, and makes possible a shift in the maintenance and monitoring philosophy from a time-based initiator, to a condition based method, relying upon automatically measured values provided by the equipment itself. By establishing a logic diagram between observed operating parameters and a corresponding, predefined action, operators can be alerted to both changes in a machine's operating characteristics, while being provided with likely solutions to resolve the changed parameter.

Electromagnetic Interference existing in an Electromagnetic environment and potentially identified as a noise source can degrade the performance of the electronic system in their vicinity. It is interesting to note that such noise is also prevailing naturally and the sources of it are given below. Unfortunately a ship out at Sea is exposed to all such sources.

Table 1.1 Classifications of natural EMI sources

A lot of man made noise is also generated and again most of this generating sources are available on operating ships.	
EMI sources due to the power network and its equipment:	
Switching operations	Static and rotary connectors
Power faults	Rectifiers
Electric motors	Contactors
EMI sources due to commercial equipment:	
Experimental furnaces	Fluorescent lamps
Boilers – Fans, Firing Equipments.	Neon lamps
Air conditioning	Medical equipment
EMI sources due to machines and tools:	
Workshop machines	Rotary saws
Rolling mills	Compressors
Welding machines	Ultrasonic cleaners

EMI sources due to communication systems:	
Radio broadcast stations on board	Citizens-band
TV Stations on board	Mobile telephones
Radar	Remote control
	Door-opening transmitters
EMI sources due to consumer devices:	
Microwave ovens	Vacuum cleaners
Refrigerators/ freezers	Hair dryers
Thermostats	Shavers
Mixers	Light dimmers
Washing machines	Personal computers

Table 1.2 Classification of man-made EMI sources

An Electronic device is vulnerable to the EMI effects caused by Man-Made noise sources in its vicinity if EMC considerations are not given. To have a rough idea of the various types of frequency and noise levels available from interference sources on ships, the following table is given:

Source type	*Comments*
Power mains disturbances	Double exponential transients with rise times of about 1 ms, fall times of tens of ms, and peak value of about 10 KV, 100 KHz- ringing waveform with 0.5 ms rising edge. Power dips up to 100 ms long. Power frequency harmonics up to 2 KHz.
Unintended radiations, switches and relays	Transients with rise times of a few nano second and levels up to 3 KV, producing frequencies into the VHF band.
Commutation motions	Frequencies up to 300 MHz at repetition rates up to 10 kHz.
Human electrostatic discharge	Rise time of 1 to 10 nano second.
Switching semiconductors	Rise time from 10 to 100 nano second at repetition rate of 1 KHz to 10 MHz for voltages up to 300 V.
Switched-mode power supplies	Continuous noise from 1 kHz to 100 MHz.
Digital logic	Continuous noise from 1 kHz to 500 MHz.
Industrial and medical equipment	Metals heating in the range 1 to 199 kHz; medical equipment operates from 13 to 40 MHz at a high power of hundreds of watts.
Intended radiations: Broadcast stations, other RF transmitters, including radar	Frequency Range 0.014 to 10000 MHz.

Table 1.3 Frequencies and noise levels from typical interference sources on board ships

1.2 BACKGROUND

This book is a study of issues, experiences, and trends in shipping system related electromagnetic compatibility (EMC). Electromagnetic compatibility and shipping systems is an area evolving from the early days of few electrical devices to the highly complex electronic components in vessels. This book will look at how the EMC of shipping systems has become a major issue, and describe the tools and techniques of shipping systems EMC. We will look at various system components architectures and EMC issues that are associated with those systems.

We'll assume the reader has a basic understanding of shipping electrical and electronic systems. Let's begin this book by defining EMC as "the ability of an electronic system to function properly in its intended electromagnetic environment, and do not contribute interference to other systems in the environment." the goal is to have the electronic system be immune from the emissions of other systems, not interfere with the operation of other systems, and not interfere with its own operation.

The basic model of EMC can be thought of as shown in Fig. 1.1. Let's assume we have device A and device B. Our goal in EMC is to have A and B operate in the presence of each other as well as operate in the presence of external environments. We do not want A to interfere with B nor do we want to B to interfere with the operation of A. We also do not want the external environment (for example, radio broadcast transmitters) to cause either A or B to operate incorrectly.

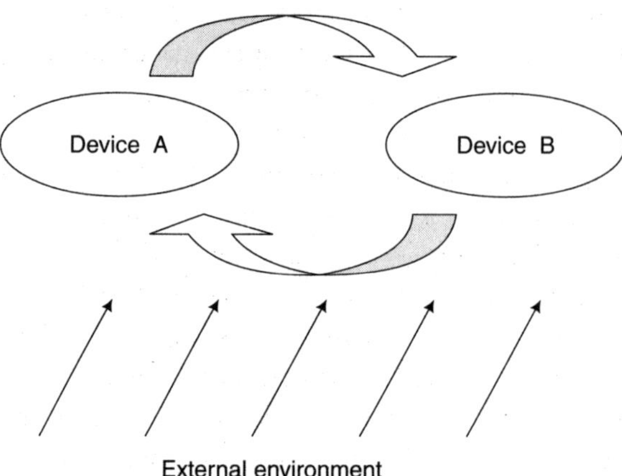

Fig. 1.1 Basic EMC model

A key concept in EMC is the "source-path-receiver" relationship, which is shown in Fig. 1.2.

Fig. 1.2 Source-path-receiver model

This shows that fundamentally, we have three basic elements comprising EMC; the first is the source, the next is the path, and the final one is a receiver. Receivers may be of two different types; "intentional" or "unintentional". An example of an intentional receiver would be a radio or television receiver, and an example of an unintentional receiver would be a computer or some type of electronic device. This is the basic model that we use in addressing EMC problems.

How can we ensure EMC with knowledge of this model?

- We could suppress the energy at the source (meaning that we could reduce the amount of energy being radiated).
- We could address the path itself; this path might be conducted through a wire, or radiated through the air.
- We could address the receiver's characteristics and make it a "hardened" receiver.

A key issue with regard to receivers of energy is the concept of immunity and susceptibility. In the shipping industry, we use the term immunity, whereas most other areas that are concerned with the EMC disciplines use the term "susceptibility". For purpose of this book we will define susceptibility and immunity related as shown in the Fig. 1.3. If we move up the vertical axis, we have increasing levels of susceptibility; and if we look to a horizontal axis and move to the right that is decreasing immunity. What we're saying is that if we have a susceptible component or system, we have little immunity; and something that is immune has little susceptibility. This is not unlike the human body immune system; if we say that we have a highly effective immune system, we are less susceptible to illness.

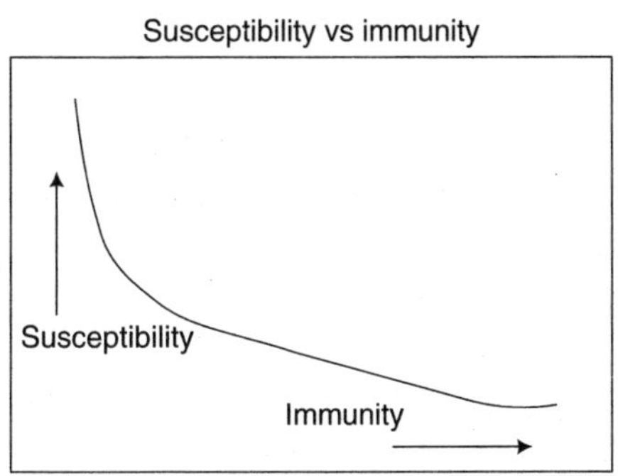

Fig. 1.3 Susceptibility and Immunity

1.3 TECHNOLOGY AND EMC

It is important to understand the origin and the migration of EMC issues which have developed as a result of technological innovations during the last century. At the beginning of the 20th century, the technology of the day with regard to communications consisted primarily of generating high frequency radio signal by generation of high-voltage discharges.

These systems were used to send short messages and allowed experimentation with propagation characteristics and high-frequency transmission, since there were few instances of EMC.

In the late 1800's to the early 1900's, there was much work done with "wireless" (the original term for radio) communication. One person heavily involved in this work was Gugliemo Marconi. Marconi did many experiments in Italy and was interested in how to send messages across the airwaves. He studied the experiments that Hertz and other pioneers in the field were performing, and he wanted to conduct such experiments himself. His work involved sending RF energy across distances using some of the techniques shown in the Fig. 1.4. The key elements in his experiments were a method to create RF energy (which was accomplished by a high voltage coil), a power source (battery), and a way to transfer the energy to the air (plates). Today antennas accomplish this energy transfer. He then constructed a way to receive the energy by creating a receiver system that consisted of plates and a device that allowed the detection of the spark energy allowing the sound to be heard in headphones. A schematic diagram of his original system is shown in Fig. 1.4. A schematic of one of his later developments is shown in Fig. 1.5.

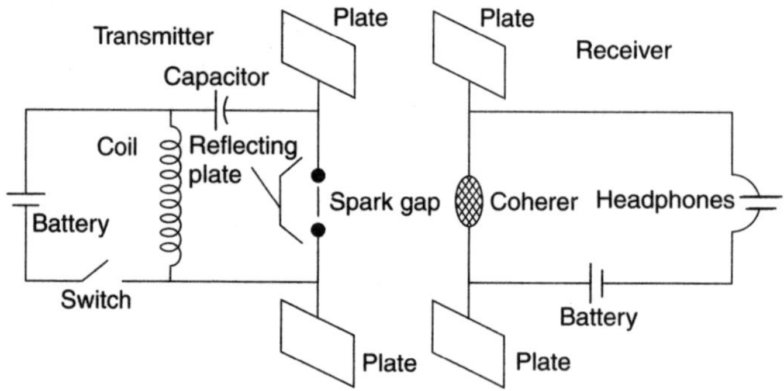

Fig. 1.4 Schematic of marconi's original wireless system

These later versions with modifications were fitted onto the earliest ships. The danger here was high voltage and compatibility with other operating equipment was relatively unknown.

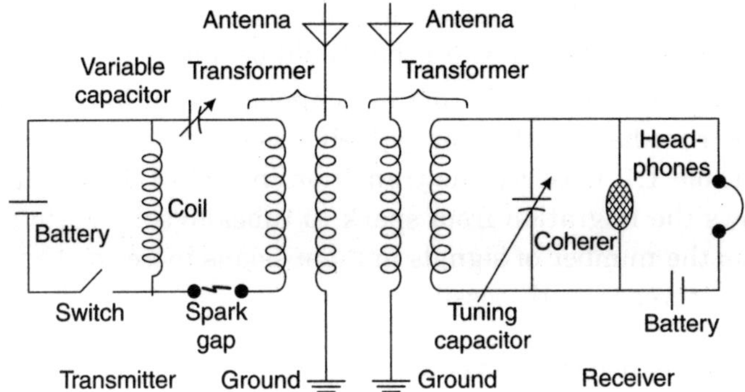

Fig. 1.5 Detailed schematic of marconi's 1900 wireless system

1.4 COMMUNICATION TECHNOLOGY EVOLUTION

If we look at the early days of electronic communication, we see that there was at first a revolution, caused by spark-gap technology, and there has subsequently been evolution. The EMC issues associated with spark technology were few because there were few receivers. Although the nature of spark-gap transmitters would be that the signals would have a wide bandwidth in general, people could identify which transmission they wanted to receive. There were minimal EMC issues.

As technology evolved and there existed a need for more complex receivers, additional components were developed. The major advance in technology was development of vacuum tubes. This was a major revolution, as it finally allowed amplification of low power circuits, impossible with passive components. Fortunately, the characteristics of vacuum tubes are that they require large amounts of power and high voltages, which minimized potential EMC issues.

The next major step in the technology evolution was "solid-state" (so called because these were solid material and did not require fragile class envelopes and vacuums), primarily developed after WWII. The first devices consisted of diodes and transistors, which were intended to replace the already many applications of vacuum valves. One characteristic of solid-state devices has actually increased EMC issues; solid-state devices operate on low-level signals, and can be affected by low-level amounts of radio frequency energy. Of course, today's ship based systems include many solid-state devices, and we've moved now to areas of high-scale integration and small geometry with even lower energy levels required to operate those devices. An additional item that has been developing recently is a high demand for FCC [Federal Communications Commission] conducts as the demand for wireless communications systems increases. This additional aspect has placed an awareness of EMC in shipping because of many communications devices on board that could be affected by EMC issues.

In summary, the combination of the technology evolution and the demand for RF spectrum has created many EMC issues that exist today. This is shown in Table 1.1. Observe the transition in table 1.3 the left-hand side where shows the migration from spark to tubes to solid-state technology, and then as the number of signals and emissions increase, the number of EMC issues increase, with possibly no end in sight.

Table 1.4 Increasing numbers of Signals Results in a Greater Number of EMC Issues

	Technology	*EMC Issues*
Revolution	Spark	Few-Not many receivers
Evolution	Vacuum tube Equipment	Some- Tubes require significant power
Transition	Solid state	Many-Low level signals, Low level operation can be affected by RF
		Spectrum utilization is high

1.5 CONVERGENCE OF TECHNOLOGY AND SHIPPING SYSTEMS

Why is EMC important in any shipping environment? In recent years, shipping systems have increased their content of both electrical and electronic devices. Electronic systems on today's vessels are very complex. For example, today's systems contain active electronics; and many electronics modules include microprocessors, transistors, and switching devices. These electronics provide increasingly more control functions on vessels. The concern is that these assemblies and components may emit energy and will also react to external sources of energy, resulting in unanticipated performance of these vessel systems.

Why study EMC? There are a number of reasons. The first one is to understand how to meet various legal requirements for EMC in different countries. For example, the European Union and Canada both have requirements for radiated emissions. In the United States, emissions from electronic devices are regulated by the FCC; however, what is unique about the shipping industry is that the devices on shipping systems are exempt from the radiated emission requirements included in "Part 15" of the rules and regulations, provided they do not cause "harmful interference."

Another reason might be meeting product requirements. Clearly ship owners and operators now are very demanding regarding system and component performance. Safety cost of retrofitting "fixes' and legal issues are also incentives to ensure EMC.

Today, although many of the components that were used on early ships (high voltage distributors) may still be used, we now have extensive use of digital devices on all kinds of systems including some very complex shipping systems. We may recall from Fourier series analysis that one of the aspects of digital devices (which operate at square waves) is that a square wave contains many harmonics. These harmonics are based upon the (10-90 percent) rise time and the fall time of the square wave signal as shown in Figure 1.6 and 1.7. We will discuss later in detail the EMC challenges of these variations in the square wave signal.

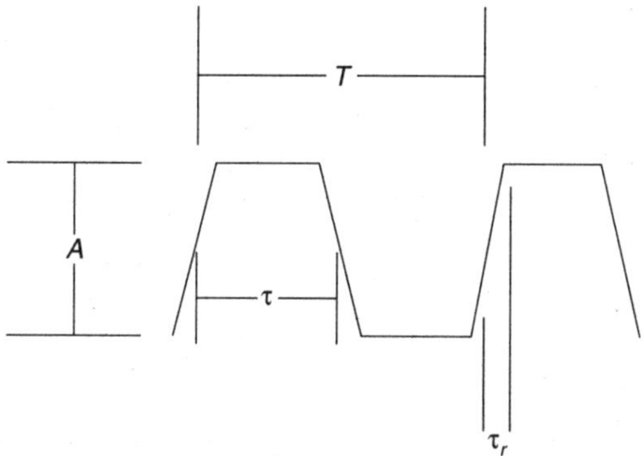

Fig. 1.6 Square wave

As shown in Fig. 1.7, we have a frequency content in the rise and fall edges that is much greater than the fundamental frequency, and a frequency content of zero in the areas where the signal is at a level value. Digital devices are used on many systems today and we will use more in future applications. This is one of the major sources of EMC (specifically emissions) that we experience in today's shipping industry, since such devices are used in machinery spaces, enclosed spaces, the navigating bridge, deck enclosures, steering spaces. In effect no machinery, component or person on board a ship today is devoid of EMR. Many ships even today may have certified equipment but not all equipment is free of EMR and no EMR / EMC mapping is done on board ships even today.

1.6 FUTURE TRENDS IN EMC MANAGEMENT

The importance of EMC to product design and manufacturing for ship board equipment will probably increase due to a number of reasons.

- In the area of computer technologies, we have seen that the clock speeds are moving higher and higher. The challenge with this technology is that as the speeds increase, cases, structures, enclosures, etc... start to have minimal effect with respect to their shielding capabilities.

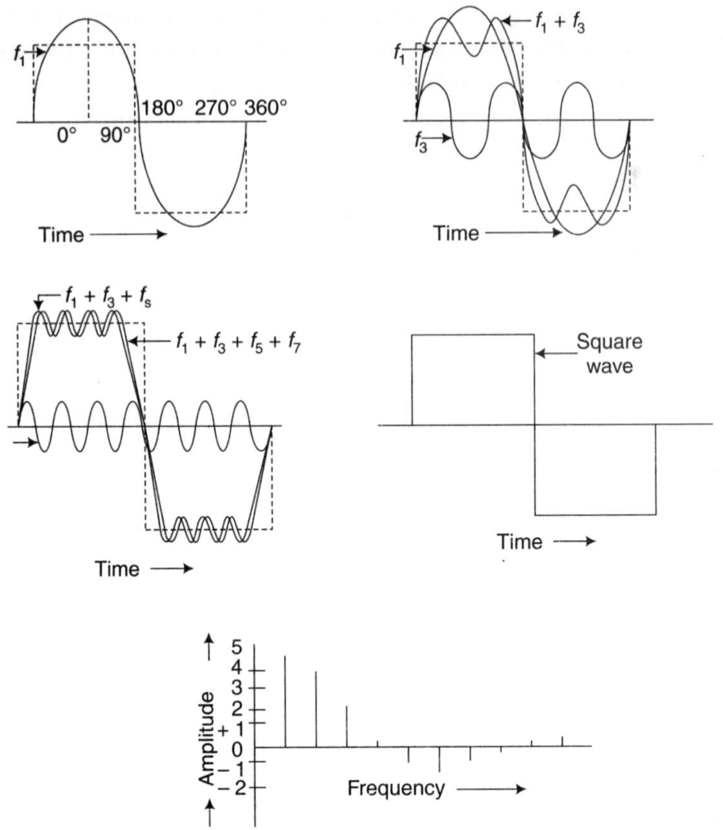

Fig. 1.7 Contribution of square wave to various frequency spectra

They can even act as fairly efficient radiators, and indiscriminate use on board ships can only create further EMC problems.

- There is also much work being done in the area of extremely short wavelength communication for high-speed wireless digital communications. The EMC aspects of this technology will need to be characterized, as EMC problems will be specifically predominant in channels, rig areas, ports, vessel cluster areas etc.,
- One is due to utilization, such as "Ultra Wide Band" (UWB). This is a low power and short-range radio system that will operate at approximately 2-3 GHz. The system is designed to be robust interference, the levels of robustness that will be needed once there is significant implementation of the systems, will be evident.
- From an antenna perspective, an effective radiator is a one-quarter or a one-half-wavelength. At 600 MHz, a one half wavelength is only 24 cm and this may require a better design for emission suppression and immunity. In other words, there is no shielding attenuation of a

lot or gap when it is half wavelength long; and once an emission from the product gets through a slot, it acts like any other wave in both its near-and far-field effects on surrounding devices. These are evident in some low range equipment fitted on fishing vessels, coastal barges etc.,

- The slot is a receiving aperture or antenna and lets in RFI environmental noise. These conditions may cause a focus on EMC at the printed-circuit-board level, in addition to the EMC issues that may been seen at lower frequencies due to wire harnesses or assemblies. Combined with the design concerns at these high frequencies, it is likely that EMC measurement difficulties will increase due to a number of reasons that include:
- The noise floor issues in spectrum analyzers and receivers may have an impact upon attempts to measure low level emissions. Antenna-to-receiver cables become closer, affecting the ability to measure signals above the noise floor of the spectrum analyzer or receiver.
- Immunity to radiated fields is more difficult because of the narrow beam width and the product's apertures, such as cable ports and slots.
- At high frequencies, the product emissions beams are narrower and are "directional", which means they are difficult to find.
- The antennas that would be used to measure emissions would also have narrower beam widths, making them difficult to aim at the product and measure the radiated signal.

Given the continual evolution in radio frequency based systems and for the examples just given, it should be seen that EMC will be a concern for many years to come!

Chapter 2

Basic Concepts Used in EMC

2.0 SOURCES AND CONDITIONS OF EXPOSURE

2.1 NATURAL BACKGROUND SOURCES

Microwave and RF radiation occurs naturally, but the intensity of natural radiation in the range of 100 kHz-300 GHz is very low in comparison with the overall intensity of man-made radiation in this range, as shown in Fig. 2.1. The intensity of natural fields is mostly due to atmospheric electricity, which is static and has an electric field intensity of about 100 V/m. This is known as the earth's electric and magnetic field. Radio emissions of the sun and stars, which are equivalent to about 10 pW/cm^2 in the range of 100 KHz to 300 GHz, also contribute to natural radiation. Local disturbances leading to increased field intensities occur during thunderstorms. Electromagnetic fields with a very wide frequency range are created (atmospheric noise) with a maximum field intensity at about 10 KHz.

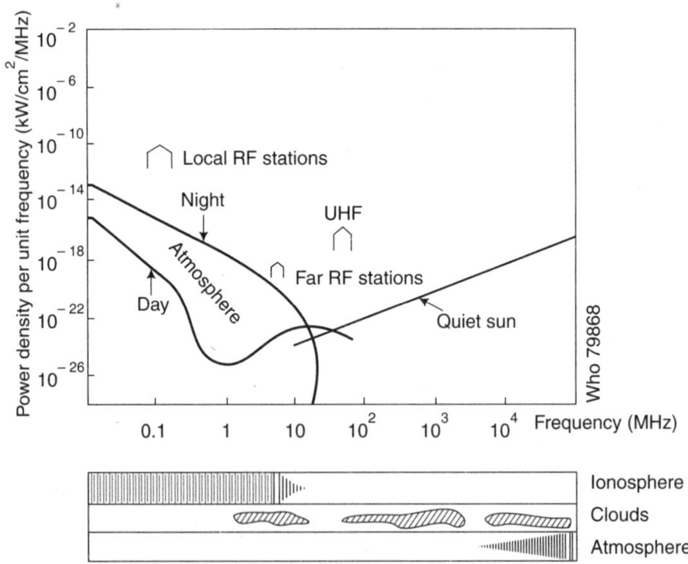

Fig 2.1 Comparison of natural background and man-made RF and microwave radiation

Artificial microwave and RF radiation constitutes a very recent environmental factor, dating back only a few decades. Depending on the frequency range, exposure from man-made sources of microwave and RF radiation may be many orders of magnitude higher than that from natural radiation and man as a species has had no opportunity to adapt to microwave and RF radiation at such environmental levels

There exists a great diversity of man-made sources, both in respect of power output and the power densities that are generated, and the frequency range in which the sources operate. According to the use of the source, different segments of the general population are exposed in different ways. There are obvious differences, depending on the development of the country, between the average exposure of the general population, the exposure of inhabitants of urban industrialized areas, and the exposure of inhabitants of rural areas.

There is also a risk of exposure to microwaves and RF in some occupations. In view of this, the discussion of man-made sources must include a description of exposure situations.

2.2 MAN-MADE SOURCES

Any appliance that generates electricity or is driven by an electric current generates electromagnetic fields. These propagate through space in the form of electromagnetic waves. Man-made microwave and RF sources may be broadly divided into 2 classes, i.e., deliberate emitters, and sources of unintentional, incidental radiation.

2.3 DELIBERATE EMITTERS

Deliberate emitters generally have a radiating element (antenna) designed to emit electromagnetic waves into the surrounding environment in a specified manner. The frequency, direction of propagation, and the point of origin are determined by the intended use of the equipment. Because of physical laws, and, in spite of the degree of perfection of the design of a deliberate emitter, some unintentional leakage, or stray radiation is always generated. This should be taken into account when evaluating a deliberate emitter as a radiation source. Typical examples are thyristors and power transformers in the panel board of ships

Unintentional radiation may occur in the form of broad-band noise or may be generated as discrete harmonics. In some instances, it is generated by sources that emit radiation outside the microwave and RF ranges. For example, while the intentional radiation of fluorescent light tubes lies in the visible light range, such tubes also generate very low levels of microwave and RF white noise.

Typical deliberate emitters include radio broadcasting and television stations, radar installations, and electronic wireless communication systems. These sources can be classified in different ways and classifications may vary from country to country depending on attitudes towards possible environmental and health effects. When classified according to the nominal power output or the effective radiated power (ERP), such emitters may be divided into high, medium, and low power sources. Radar systems used for tracking and guiding purposes, as well as sources used in satellite systems are among the most powerful. It was reported as early as in 70's that in an advanced country, that there were 20nonpulsed unclassified sources with average effective radiated power (ERP) ranging from 5 GW to 31.6 GW and one experimental source with an average ERP of 3.2 TW (3.2×10^{12}W). All these sources were used in conjunction with satellite systems. A further 144sources had an average ERP of 1 MW or more. The twenty most powerful unclassified pulsed (radar) sources had average ERPs between 8.7 MW and 840 MW and peak ERPs ranging from 35.4GW to 2.8 TW; 229 unclassified pulsed sources had peak ERPs of 10 GW or more. This may be compared with television or amplitude modulation (AM) broadcasting stations in which the power of the transmitters is of the order of tens of kW (usually about 50 kW) or radiotelephones (walkie-talkies), in which ERPs may be of the order of a few watts or less. In 2004 the nos. of emitters has increased 100 fold and by 2010 it shall be a 1000 times more in numbers.

Another approach towards classification of sources is to examine the configuration of the radiated fields and their propagation in space. Directional radiating elements (antennae) generating intense focused beams and multidirectional, variously polarized antennae may be used. Taking into account the power of the transmitter and the type of the radiating element, the magnitude of distances (or zones) at which various intensities of radiation (power densities on strength of E or H fields) occur can be computed. In this case, the classification of sources also depends on arbitrarily chosen levels of radiation intensity. This approach may be useful in the siting of sources and in establishing "safe", "hazardous", and "danger" zones around a source, with advocate warning signs specially on ships with higher such risks, so that precautions can be taken.

Deliberate emitters may be also classified according to the mode of generation. Microwaves and RF may be generated continuously or in pulses and both continuous and pulsed wave generators may operate for long periods (up to 24 h per day) or short intermittent periods. The generated radio signal may be frequency, amplitude, or pulse-modulated. Sources with moving directional antennae and sources generating mobile narrow beams may illuminate a point in space intermittently with a time-varying intensity ranging from zero to extremely high, at pulse peak power. Because of these complexities and since a point in space may be illuminated by radiation originating from several sources, the determination of the total or average quantity of energy delivered at this point during a period of time, may be difficult and may necessitate the use of sophisticated equipment and advanced computing methods.

2.4 OTHER PHYSICAL CONSIDERATIONS

A detailed analysis and interpretation of the perturbing effects of objects placed in the path of microwave or radio frequency beams requires the solution of Maxwell's field equations for the appropriate boundary conditions. However, an important insight can be obtained by comparison with the shorter wavelength visible radiation, to which the same equations apply. The general laws of geometrical and physical optics remain valid: particularly the latter because of the wavelengths involved and because deliberate generators of microwaves and RF emit coherent radiation, i.e., the wave fronts are regular and radiated over a narrow band of frequencies at any one time.

A set of four fundamental equations describe all electric and magnetic fields. Their solution for real materials requires knowledge of the macroscopic electrical and magnetic properties of the materials reflected into the path of the incident beam will form standing waves and, at a distance of a few wavelengths, diffraction effects can cause additive and subtractive interference. Both effects have been convincingly demonstrated at 1 GHz with man as the perturbing influence. Diffraction and internal reflections can also take place when radiation penetrates heterogeneous materials such as body tissues, leading to markedly no uniform internal fields and energy deposition. As in geometric optics, the combination of a high refractive index and convex body contours behaves like a strong convex lens focusing the penetrating radiation causing radiation to penetrate and harm tissues. Absorption and internal scattering within the body will limit the extent of these effects. Without the absorption of energy to initiate some change, there cannot be any biological effects.

Direct radiation is usually polarized, i.e., the E and H fields are oriented parallel to particular orthogonal planes or rotate in an ordered fashion. The plane of polarization of the reflected radiation will, thus, be changed, complicating the measurement of the combined beams and investigation of the biological effects.

Orientation with respect to the plane of polarization is important in some measuring instruments and in the distribution and total energy absorption in animals and in man. In some radar applications, typical equipment may emit pulses of 1 microsecond (μs) with a pause between pulses of 1 millisecond (ms). This constitutes a factor of 10^3 in the values of instantaneous power radiated and deposited, and of 30 in the electric fields, compared with continuous wave (cw) generation at the same average power. Thus, instruments to be used in pulsed fields must have a wider dynamic range and more robust burn-out characteristics.

2.5 AIRBORNE/SHIPBOARD COMMUNICATIONS AND NAVIGATION

RF over Fiber technology is becoming commonplace on naval ships and aircrafts. Analog fiber optics allows high bandwidth, high frequency global positioning systems (GPS), microwave transmission, radio communications, and radar circuits to be transported seamlessly throughout the vessel or airborne platform. Besides the advantages of greatly increased bandwidth, network design flexibility and capacity, fiber optics also saves cost, space and weight-in some cases tons of weight on large, military or commercial aircraft and ship borne platforms. In military and commercial applications, direct benefits of lightweight designs include increased payloads and fuel efficiencies plus reduced mission/flight time.

Fiber optic links are most attractive for reducing the bulk and weight of microwave links aboard ships, i.e. heavy cabling and wave guides can be eliminated between the control towers, antenna arms and the bridge. Additionally, these rugged, robust links provide redundancy in case of attack or weather related incidents. Another benefit is the reduction in RF emissions from leaky coaxial cable.

2.6 ANTENNAS

Many EMC issues result from energy that is transferred by radiation from a source. In order to understand this radiation of energy, it is useful to refer to some basic electromagnetic principles. One of these principles is the "isotropic point radiator" of energy. As this point source has zero radius and radiates equally well in all directions. This is shown in Fig. 2.2.

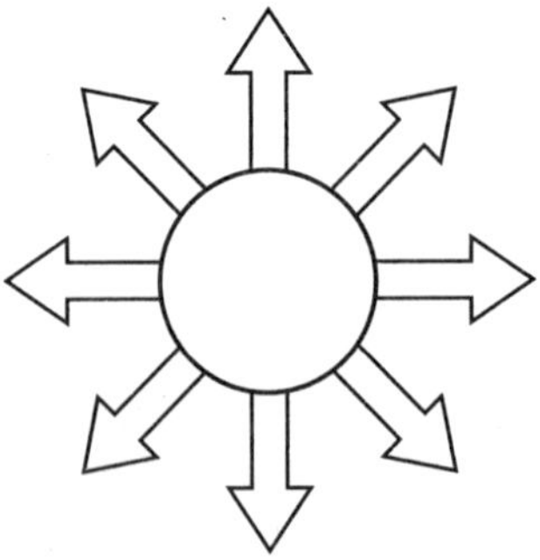

Fig. 2.2 Isotropic radiator

Real energy sources that intentionally transfer energy by radiation are called "antennas" and have several key characteristics which differentiate them from isotropic radiators. The first is directly, which is the direction of the maximum energy transfer. The second is gain, which relates to the shape of the energy transfer pattern.

If we look at the directivity of an antenna, it is essentially "the map of the gain". Gain refers to the ratio of any portion of the pattern to any other portion. In EMC work, another issue is antenna factor, which relates to the transfer function between energy and voltage at the terminals.

We will now discuss basic ship board antenna concepts and designs. This subject of the physics and mathematics behind antennas can be complicated and time consuming. There are numerous references on antennas that the reader is encouraged to review for detailed understanding. The intention in this text is to review basic antennas that may contribute to or create EMC problems.

Two common types of antennas used on board ships are "quarter wave" and "half wave" antennas. These names refer to the fact that their physical dimensions approximate a portion of the wavelength, which is determined from the speed of propagation and the frequency of intended operation (discussed previously). For example:

A half-wave antenna used to receive a signal at 100 MHz would be approximately 1.5 m long. An element of a quarter-wave antenna for the same frequency would be approximately 0.75 m long.

These antennas radiate with a maximum in directions 90 degrees from the axis of the elements. Consequently, these antennas are referred to as "omni-directional" antennas.

Another basic types of antenna are the "gain" antenna mostly used in shore stations. This antenna differs from an omni-directional antenna in that this antenna both transmits and receives energy primarily from certain directions. In some ways both a half-wave and a quarter-wave antenna exhibit some degree of directionality. While typically not defined as gain antennas, they do have characteristics that make them sensitive in certain directions, as shown in Fig. 2.3.

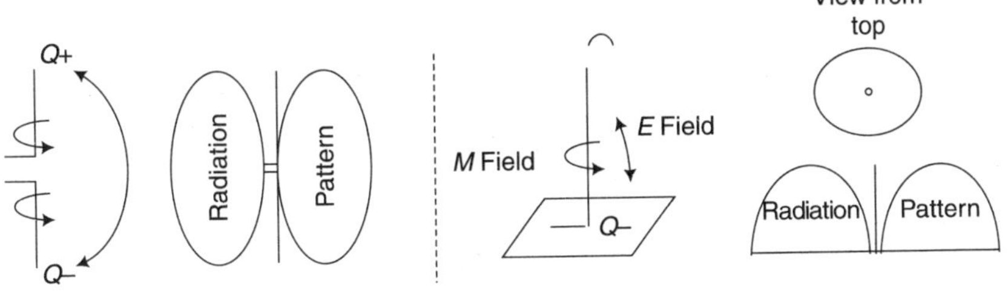

Fig. 2.3 Typical antenna patterns for half wave dipole (Left) and quarter wave vertical (Right)

In addition to directionally, another characteristic of these antennas is impedance at resonance (the radiation resistance). Radiation resistance means the effective resistance that the antenna exhibits when connected to a source. A half-wave antenna commonly used as a dipole antenna has a radiation resistance of approximately 73 ohms. A quarter-wave antenna, typically used with a counterpoise surface (generally called by many a "ground plane") has a radiation resistance of approximately 37 ohms.

Let's look in more detail at the dipole and quarter wave antennas. Dipole antennas are typically constructed horizontal to the ground and for communication purposes, are ideally located several wavelengths (at the frequency of operation) above the deck level. Quarter-wavelength antennas are typically mounted with their main radiating element located vertically to the deck level, and have one or more radials parallel with the ground. This is termed a ground-plane antenna because the radials approximate the earth itself. More correctly, the radials are the "counterpoise" for the antenna, and create an "image" element. This antenna was developed to meet the need for an efficient, inexpensive base station antenna for use in communicating with mobile units.

Figure 2.3 shows half-wave dipole and quarter-wave vertical antenna patterns.

2.7 OMNI DIRECTIONAL ANTENNA

2.7.1 Quarter-Wave Vertical

What are the dimensions of typical quarter-wave antennas that are commonly used for mobile communications to use when determining the length of quarter-wave antennas? If you recall that the calculation for wavelength is equal to the speed of propagation (which in free space is 300 million meters per second) divided by the frequency in MHz, then the length of the vertical element of the quarter-wave antenna would be the wavelength divided by four. Table 2.1 shows frequencies common vertical antennas used for ship board mobile communications and their approximate length.

Table 2.1 Frequency, Wavelength, and Quarter-Wavelength

Frequency, MHz Meters	*Wavelength, Meters*	*Quarter-Wavelength, Feet or inches*
27 (CB)	11.00	27.00 (9 Feet)
45 (Land Mobile)	6.70	1.70 (5.5 Feet)
150 (Land Mobile)	2.00	0.50 (19 Inches)
850 (Cellular Telephone)	0.35	0.09 (3.5 Inches)

1 meter = 39.37 inches

2.7.2 Ground Plane

If we use the term ground plane for an antenna type, one meaning could be as shown in Fig. 2.4, where we would have the vertical element over the reference or the "ground". As shown in Fig. 2.5 the ground looks like the image of the vertical element or the counterpoise, which then resembles a one-half-wave dipole. The difficulty when referring to these types of antennas can be seen in this example; if we have a vertical element on a ship as in figure 2.5, where is the ground plane?

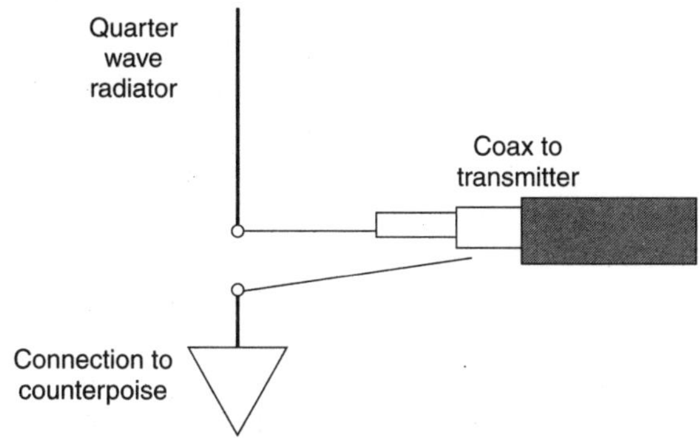

Fig. 2.4 Basic representation of a simple vertical antenna

A quarter-wave perpendicular to a reflecting plane is electrically the same as a half-wave dipole.

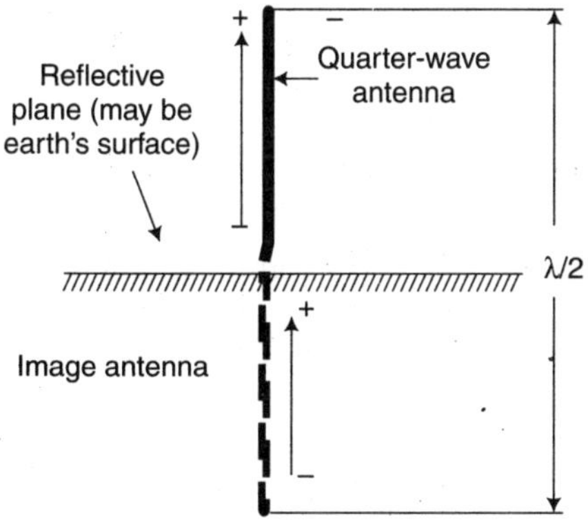

Fig. 2.5 Ground plane antenna operates becasue of the creation of an "Image"

2.8 OTHER ANTENNA TYPES

2.8.1 Antenna Arrays

An antenna array is a multiple of active antennas coupled to a common source or load to produce a directive radiation pattern. Usually the spatial relationship also contributes to the directivity of the antenna. Use of the term "active antennas" is intended to describe elements whose energy output is modified due to the presence of a source of energy in the element (other than the mere signal energy which passes through the circuit) or an element in which the energy output from a source of energy is controlled by the signal input. The relative amplitudes , constructive and destructive interference effects among the signals radiated by the individual antennas determine the effective radiation pattern of the array. A phased array may be used to point a fixed radiation pattern, or to scan rapidly in azimuth or elevation.

Phased array radar systems are also used by warships of several navies including the Chinese, Japanese, Norwegian, Spanish, Korean and United States' navies in the Aegis combat system. Phased array radars allow a warship to use one radar system for surface detection and tracking (finding ships), air detection and tracking (finding aircraft and missiles) and missile uplink capabilities. Prior to using these systems, each surface-to-air missile in flight required a dedicated fire-control radar, which meant that ships could only engage a small number of simultaneous targets. Phased array systems can be used to control missiles during the mid-course phase of the missile's flight. During the terminal portion of the flight, continuous-wave fire control directors provide the final guidance to the target. Because the radar beam is electronically steered, phased array systems can direct radar beams fast enough to maintain a fire control quality track on many targets simultaneously while also controlling several in-flight missiles.

2.8.2 Unanticipated Antenna

In addition to intentionally creating antennas, connecting conductors to components creates a system that did not exist when considering only the components. The contribution of the conductor results in increased efficiency of energy transfer and behaves like an antenna at lower frequencies than would be possible with just the component itself, which is a smaller size than the combination of the conductor and the component. Empirical data suggest that a conductor longer than 10 percent of a particular wavelength starts to become an efficient radiator. For example, a printed circuit board trace with a length as short as approximately 0.15m (6 inches) could be an efficient radiator of the system emissions at approximately 200 MHz! The ten percent rule is reasonable, since a quarter wave antennas is an excellent radiator.

2.8.3 Reduced Size Antennas

Another type of antenna can be created by incorporating a series inductive element in a conductor that is physically shorter than the electrical length of quarter-wavelength antenna. By use of this inductive element, the antenna can be physically shorter then one-quarter of the wavelength at the frequency of use, yet be at a quarter wavelengths electrically. This principle is used on vessels that utilize HF frequencies for communications. Without the use of the inductive element, the antennas would be very long and impractical on a vessel. The reason that this antenna can utilize an inductive element is that the addition of the inductive element makes the antenna appear "electrically" one-quarter of the wavelength, and therefore is closely matched to the transmitter impedance of 50 Ohms.

2.8.4 Gain Antenna

Another common type of gain antenna used in both communication systems(or radio and television reception) as well as EMC work is titled the Yagi. These_antennas are sometimes referred to as "Fishbone" antennas. In general, these types of gain antennas consist of a number of elements perpendicular to the antenna boom, used to support the elements.

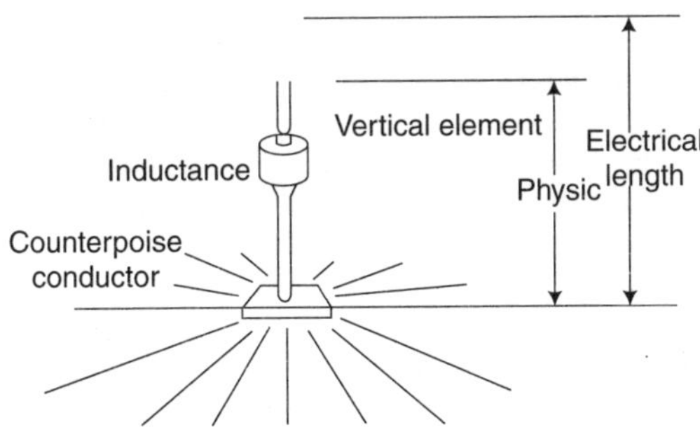

Fig. 2.6 Center loaded vertical antenna

2.9 OTHER COMPONENTS IN EMC

We will discuss basics of other components used in EMC work. It is important to understand the physics of each of these components, as they comprise a large portion of the intended and unintended current paths in EMC. We shall discuss the following in breif:

- Inductors
- Capacitors
- Resistors
- Cables and Transmission Lines
- Shields

2.10 IDEAL AND ACTUAL COMPONENTS – CAPACITORS AND INDUCTORS

When working on EMC issues it is important to remember that we are dealing with actual components, unlike textbook theoretical components. For example, inductors have only inductance, capacitors have only capacitance, resistors have only resistance, etc.

A key question to ask at this point is how many products are built from only theoretical components? The answer is simple-none! So another issue in EMC work to comprehend is that there is a difference between ideal components and actual components. The difference between an ideal and an actual capacitor is shown in Fig. 2.7(a), and the ideal vs. actual inductor in Fig. 2.7(b). The ideal inductor has only inductance, and actually the real inductor has parallel parasitic capacitance, as shown. The ideal capacitor has only capacitance, while the real capacitor has series inductance as well.

(a) Inductor's equivalent circuit

C (Stray capacitance)

C (Stray capacitance)

At high frequencies, the inductors is dominant

At high frequencies, the stray capacitance is dominant

(b) Effect of stray capacitance

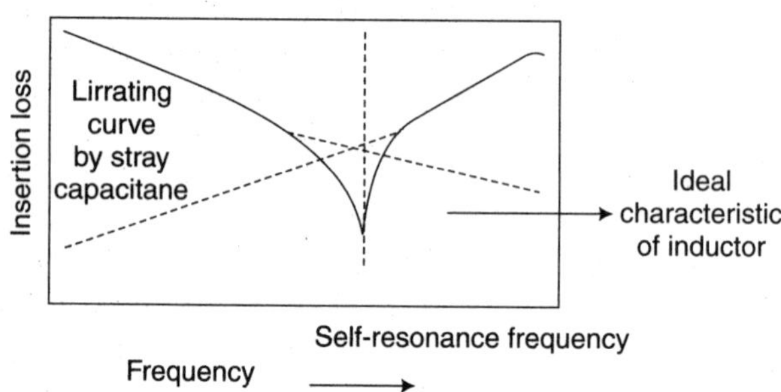

Fig. 2.7 Ideal Vs actual inductor

A refresher about capacitance follows:

- Capacitance is Proportional to Area of plates, and Dielectric Constant.

- Capacitance is inversely proportional to the distance between the plates.
- At low frequencies, the capacitance is dominant.
- At high frequencies, the inductance is dominant.

Because of self-resonance effects, certain types of capacitors are preferred for use at specific frequencies, as shown in Table 2.2.

Table 2.2 Optimum Frequency Range for Different Types of Capacitors

Type	*Frequency range*		
Aluminum Electrolytic	1 Hz	-	10 kHz
Tantalum Electrolytic	1 Hz	-	10 kHz
Paper or Mylar	100 Hz	-	5 Mhz
Ceramic	1 kHz	-	100 MHz
Plastic Film	1 kHz	-	9 GHz
Mica, Glass or Ceramic	5 kHz	-	10 GHz

The issue here is what type of capacitor material has been used even in components of reputed manufacturers leave aside local or indigenously assembled equipment. The wrong use of a capacitor, unsuited for a specific frequency range, even though operationally adequate may give erroneous parasitic effects.

Different types of inductors have different characteristics, as shown in Table 2.3. Leakage flux should be minimized to avoid coupling to near by circuits.

Table 2.3 Inductor Characteristics

Inductor type	*Leakage flux*
Open Core	High
Closed Core	Low
Toroid	Very Low

A third fundamental component is the resistor. Resistor types are summarized in Table 2.4.

Table 2.4 Resistor Characteristics

Resistor type	*Cost*	*Frequency sensitivity*	*Current handling capacity*
Carbon	Low	Low	Low
Wire Wound	High	Inductive	High
Thin Film	Medium	Moderately Inductive	Low
Chip	Medium	Moderately Inductive	Very Low

Depending on the type of construction, resistor may also exhibit frequency sensitivity and self-resonance. Check the manufacturer's specification sheets to ensure that the impedance of the resistor is not affected by the signal frequencies.

2.11 TRANSMISSION LINE

A transmission line is used to transfer energy from the antenna to a transmitter. See Fig. 2.8.

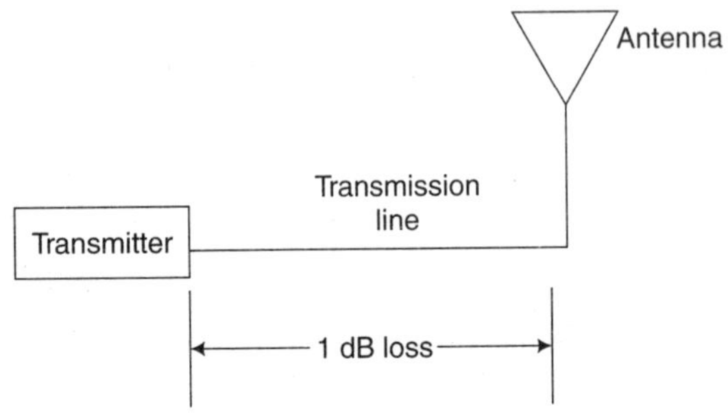

Fig. 2.8 A transmission line connects the transmitter to the antenna

There are several types of transmission lines and we will discuss the most common ones here. They all have similar characteristics and functions; however, the construction can be very different and each type can have very different requirements. Fig. 2.9 and 2.10 shows the two primary types of transmission lines, "open wire line" and "coaxial line. There are important differences between these two types of transmission lines. The first one that can be seen is that the open wire line in some ways resembles a ladder, with the conductors located opposite from one another and the insulator between them. This transmission line has some positive characteristics as well as some disadvantages. One advantage of this type of transmission line is that connections to it can be made very easily; no special connectors are required. Another advantage is that it is fairly inexpensive, and exhibits low loss. The disadvantages are that because of its construction, it provides very little shielding from being radiated from the transmission line. Coaxial cable compensates for some of these advantages; however, it is typically more expensive and does require special connections usually consisting of cylindrical connectors that must be fastened to the coax so it can be connected to the mating connector. Another disadvantage with coax is higher loss, resulting in less of the power being transferred from the antenna to receiver or from transmitter to the antenna. Also, the wave propagation speed is less than the propagation in free space, which means that the wavelength is different within the cable than in free space. The propagation velocity in the cable is less than in free space; therefore, the wavelength in the cable is somewhat shorter.

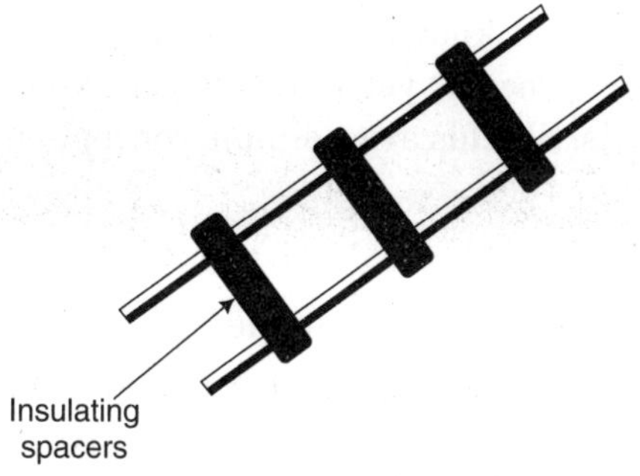

Fig. 2.9 Open wire transmission line

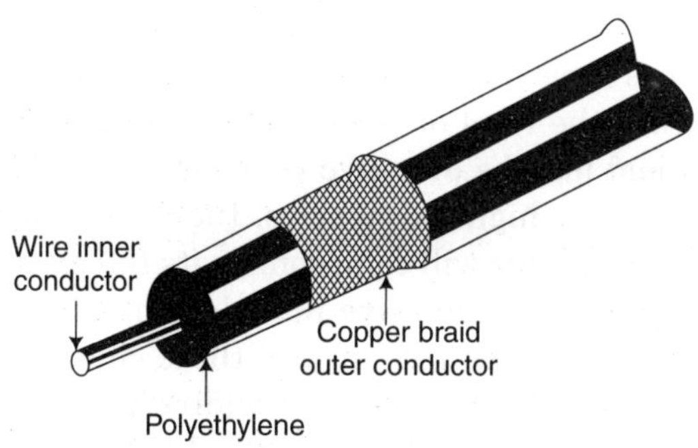

Fig. 2.10 Coaxial transmission line

2.12 CHARACTERISTICS OF COMMONLY USED TRANS MISSION LINES

There are many types of coaxial cables, many with similar impedance, although they may vary in terms of their maximum power handling capability. The maximum power handling capability is a function of the dielectric it can withstand. Another key item is the outside diameter (OD) of the cable. It can be seen that, in general, as the OD increases, so does the maximum voltage rating. There are several types of dielectric material that are used, including air, in the cable. The other type of cable shown is the parallel line or the "open wire" line. The impedance of the cable is a function of the distributed inductance and capacitance. It can be seen that, as the conductors are moved apart, the capacitance decreases, and the maximum operating voltage that can be applied increases, and the impedance also increases.

2.13 SHIELDS

2.13.1 Purpose of Shields

A subject that is not well known with respect to the mathematical and quantitative analyses process, yet is intuitively known is the concept of shielding for EM fields. Shields are a simple concept and in EMC serve two functions:

- Keep emissions inside
- Keep external source/energy outside

As most people are aware, a role of a shield is to isolate fields from components that are operating. Shields can be used to provide protection against both types of EM fields, electric and magnetic. Recall that voltage causes electric fields and current causes magnetic fields. A metric used in shielding work is the measure of how well the shield does its job. This is expressed in terms of shielding effectiveness, abbreviated as SE. So it is measured in dB and is used to measure the amount of isolation to field strength levels. It is typical for good shields to provide 100 or more dB of SE. This means that the level of the field strength (for electric fields) on one side of the shield may be over five times to achieve these high levels of SE, especially at very high frequencies, the shield must not have any holes, slots, or openings that will compromise its integrity. A test chamber for EMC may need to have extensive inter locking pieces to provide this level of SE, and is part of the reason why these chambers are expensive to acquire and maintain. A key type of shielding can be witnessed in the portable chambers for testing transmission for EPIRBS in Life Raft Service stations. Another complication, which is being experienced more and more frequently because of the increase in speed for digital computers, is that openings are required in the enclosure to attach peripherals, power cables, and I/O devices. Even small openings for those functions can compromise the effect of the radiation shielding.

2.13.2 Shielding Effectiveness

Let's examine the metrics and calculations involved in the determination of shielding effectiveness.

The SE would be the following:

$$SE = 20 \log (10/0.3) = 29.6 \text{ dB}$$

As was stated earlier, SE is expressed in dB and can be calculated as the ratio of the field strength on one side of the shield and the field strength on the other side of the field. For example see Fig. 2.11 and Fig. 2.12

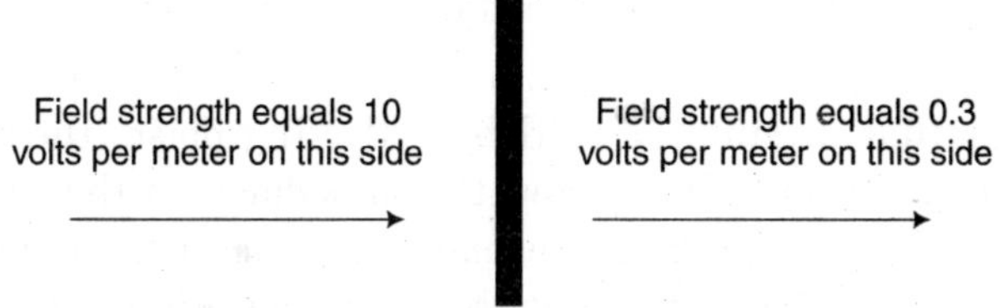

Fig. 2.11 Shielding effectiveness

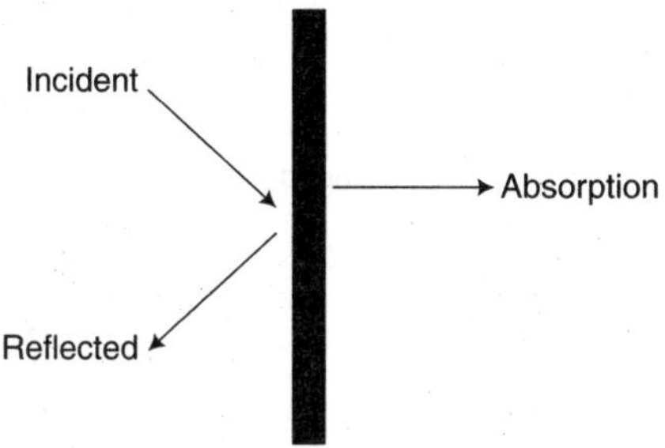

Fig. 2.12 Reflected wave from shield

What this means is that there is a 30 dB reduction of field strength because of the shield.

The actual process that takes place in shielding consists of two main items:

- The first is the reflection of the incident field
- The second is the absorption of the energy within the shield material.

The relative contribution of each one of the mechanisms is dependent on whether the field is electric or magnetic, and low or high frequency. This is shown in Fig. 2.12.

This means that the source of the energy is on the left side of the shield, and the device to be protected is on the right side of the shield. For electric field shielding (at low frequencies), the reflection is the primary cause of the SE, and at high frequencies, absorption of the energy occurs.

2.14 FILTERING OVERVIEW

As stated before, the purpose of the EMI filter is to prevent the entry or exit of undesired electromagnetic energy from equipment. A filter absorbs the noise energy through the use of elements such as resistors and ferrite components, or reflects the noise energy back to the source through use of reactive elements. Generally, EMI filters are low-pass filters with effectiveness depending on the impedances of the elements at either end of the filter.

For a filter that attenuates EMI by reflecting noise, the filter should provide maximum impedance is low, the impedance of the filter from the load viewpoint should be high. If the load impedance is high, the impedance of the filter from the load viewpoint should be low. Figure 2.13 gives filter configuration examples for various load and source impedances.

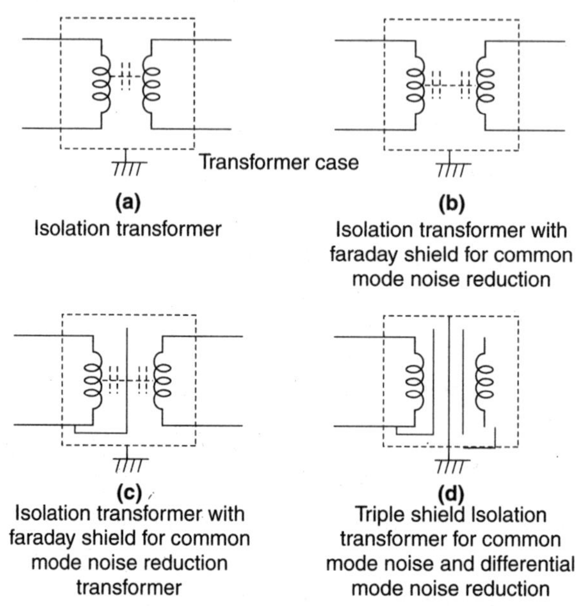

Fig. 2.13 Isolation transformer configurations

EMI filters are single-section filters or several single-section filters cascaded together for more attenuation. It has been demonstrated that a two-section filter has a lower optimum weight than a single-section filter when by design both have identical filtering properties. The number of sections and configuration are not limited to this presentation. It is important to remember to isolate the input and output cables of the filter. Isolating input and output cables from each other prevents the cables from coupling to each other and by passing the filter. Isolation may be accomplished by placing the input cables and the output cables on opposite sides of the filter. However, to properly isolate the cables and prevent noise from bypassing the filter, the filter may have to be shielded by placing it in a shielded enclosure.

2.15 ENCLOSURE SHIELDING

Most books on shielding delve into a comprehensive coverage of shielding theory that is beyond the scope of this handbook. Guidelines provided here provide a minimum of mathematics and theory. Shielding of EM fields is accomplished through reflectance or absorption of the fields by a barrier. In most applications, the barrier is a metal, although coated and conductive plastics are being used more frequently in commercial applications. An important point to remember in shielding is that the actual shielding provided by a metal barrier depends on the type electromagnetic field that predominates. Reflection is highly effective against predominately electric fields and plane waves and has little effect on predominately magnetic fields.

Absorption is the mechanism in predominately magnetic field attenuation. Reflectance increases with surface conductivity of the shield and decreases with frequency. Absorption increases with:

- Thickness of the shield
- Conductivity of the shield
- Permeability of the shield
- Frequency of the incident field

Absorption in a metal barrier is exponential in nature, i.e., as an electromagnetic field passes through a metal barrier, the amplitude of the electromagnetic field decays exponentially. At some distance into the metal barrier, the amplitude of the impinging electromagnetic field has decreased to 1/e or 33 percent of the amplitude at the surface of the barrier. The distance at which this occurs is called the skin depth of the metal. The formula for skin depth is given in equation

$$\mathbf{d = 2.6 / \sqrt{\mu r \delta r f}\ MHz\ in\ mils}$$

Where μr is the permeability of the metal relative to copper,

δr is the conductivity of the metal relative to copper, and

fMHz is the frequency of the electromagnetic field impinging on the metal.

The skin depth concept is shown in Fig. 2.15. Table 2.4 lists plane wave skin depths for copper and aluminum at various frequencies.

The performance of a shield in reducing the electromagnetic energy that passes through it is known as its shielding effectiveness. The equations below define the shielding effectiveness (in decibels) for electric fields and magnetic fields:

SEdB = 20 log 10 {Ein/Eout} for electric fields

SEdB = 20 log 10 {Hin/Hout} for magnetic fields

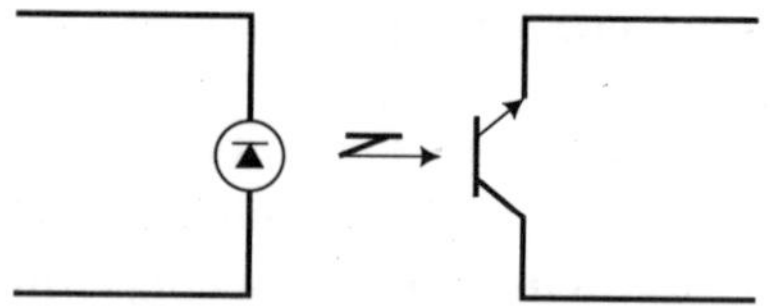

Fig. 2.14 Opto isolator schematic

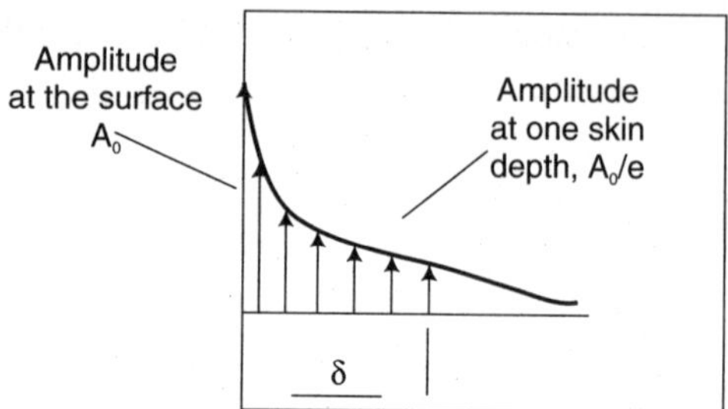

Fig. 2.15 Schematic of skin depth

Where Ein (Hin) is the field strength incident on the shield, and Eout (Hout) is the field strength after passing through the shield.

Shielding effectiveness is shown in Fig. 2.16. Note: At one skin depth the SEdB of a metal is at least 8.7 dB and at 2.3 skin depths the SEdB is at least 20 dB.

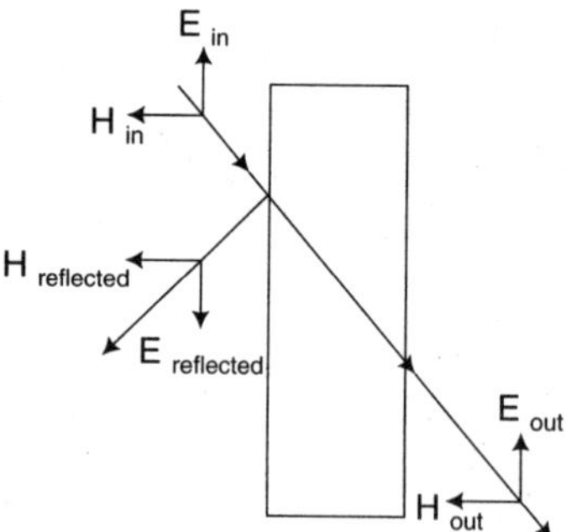

Fig. 2.16 Schematic of shielding effectiveness

The above discussion assumes that the barrier or shielding material is homogeneous and large such that there is no leakage or edge effects. The shielding effectiveness expressed in the equation is degraded by apertures for connectors, switches, and I/O lines and seams for doors, access panels, and cover plates. These apertures and seams serve as leakage paths for electromagnetic energy; this leakage lowers the SEdB of the barrier.

Finally, a few shielding rules of the thumb:

- For a predominately electric field or plane, wave; use a good conductor (copper or aluminum) to maximize reflection loss.
- For a high frequency magnetic field (frequency >500 kHz), use either a good conductor or a material with a high permeability, μr.
- For a low frequency magnetic field (10 kHz>frequency> 500 kHz), use a magnetic material such as steel, for frequency <10 kHz use a material with a high permeability, μr, to maximize absorption loss.
- Reflection loss varies with the type of field; absorption loss is independent of the field.
- A metallic shielding material thick enough to support itself usually provides good electric field shielding at all frequencies.

Table 2.5 Skin Depths at Various Frequencies.

Frequency	*d for Copper (mils)*	*d for Aluminum (mils)*
10 kHz	26	33
100 kHz	8	11
1 kHz	2.6	3
10 kHz	0.8	1
100 kHz	0.26	0.3

2.16 SHIELD DISCONTINUITIES

As stated in the previous section, the shielding effectiveness of an enclosure is degraded by the introduction of discontinuities into the enclosure. These discontinuities are the holes, seams, and joints found in nearly all electrical equipment. Leakage through seams, holes, and joints is usually a greater concern than the shielding effectiveness of the shield material. The methods presented here are equally adequate for minimizing magnetic and electric field leakage; only the types of shielding materials differ. Discontinuity rules of thumb include:

- The amount of leakage from a discontinuity depends on the maximum linear dimension of the opening and the frequency of the source.
- A slot or rectangular hole may act as a antenna when the maximum linear dimension of the slot becomes greater than 1/10 of a wavelength.
- A large number of holes allow leakage than one large hole of the same total area.
- A hole shaped to form a waveguide (the depth of the hole is greater than the diameter of the hole) can offer greater attenuation than a “regular” hole pattern for frequencies lower than the wave guide’s

critical frequency. This critical frequency is roughly the frequency at which the maximum linear dimension of the opening of the waveguide attenuation is dependent on the length of the waveguide and is called waveguide below cutoff.

- For seams and joints it is necessary to maintain a continuous metal-to-metal contact along the seam or joint to ensure shielding integrity.
- The preferred seam for preventing EMI leakage is a continuous weld.
- When bolts or rivets are used to make a bond, the shielding effectiveness depends on the number of rivets or screws per linear cm, the mating pressure at the contact surface, and the cleanliness of the two mating surfaces.
- The higher the number of rivets or screws per linear cm, the greater the shielding effectiveness.

For equipment enclosures that require ventilation, the following materials (in descending order of attenuation) should be used to cover the opening:

1. Waveguide below cutoff panels (honey-comb panels)
2. Perforated metal sheet
3. Woven or knitted metal mesh.

Chapter 3

Effects of EMI and EMC

3.1 PROPERTIES OF MICROWAVE AND RADIOFREQUENCY (RF) RADIATION

Radio waves in the frequency range 100 kHz-300 GHz are non-ionizing electromagnetic radiation, and can be described in terms of time-varying electric and magnetic fields moving though space in wavelike patterns, as represented in Fig. 3.1.

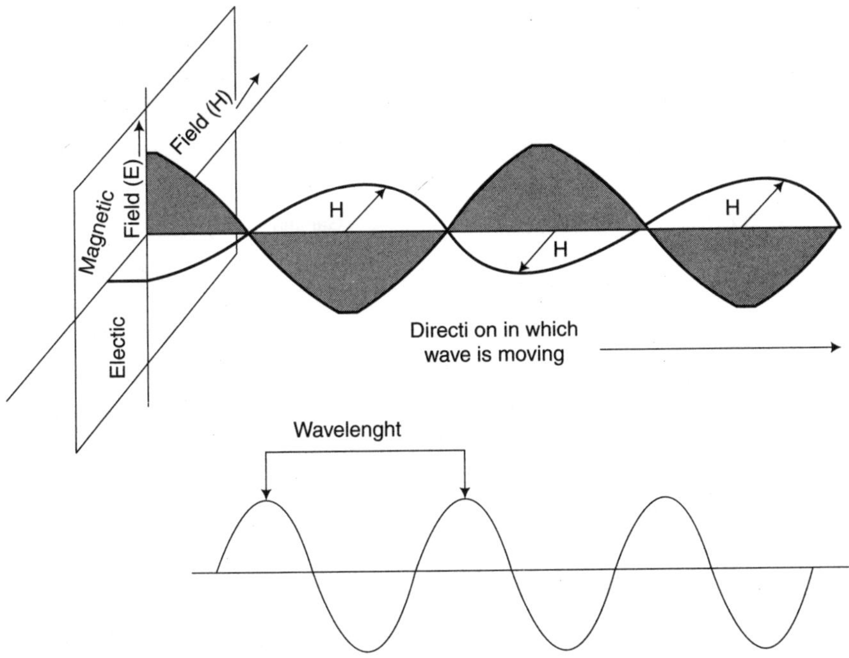

Fig. 3.1 An electromagnetic monochromatic wave. Electromagnetic waves consist of electrical and magnetic forces which move in consistent wave like pattems at right angles to one another – WHO

The wavelength (the distance between corresponding points of successive waves) and the frequency (the number of waves that pass a given point in

1 second) are related and determine the characteristics of electromagnetic radiation. The shorter the wavelength, the higher the frequency. At a given frequency, the wavelength depends on the velocity of propagation and therefore, will also depend on the properties of the medium through which the radiation passes. The wavelength normally quoted is that in a vacuum or air, the difference being insignificant. However, the wavelength can change significantly when the wave passes through other media. The linking parameter with frequency is the velocity of light (3×10^8 m/second in air). The velocity decreases and the wavelengths become correspondingly shorter, when microwaves and RF radiation enter biological media, especially those containing a large proportion of water.

Another related property of electromagnetic waves is the photon energy, which increases linearly as the frequency increases. Fig. 3.2 shows the spectrum of electromagnetic radiation ranging from highly energetic ionizing radiation with extremely high frequencies and short wavelengths to the less energetic non-ionizing radiation with the much lower frequencies and longer wavelengths of radio-frequencies.

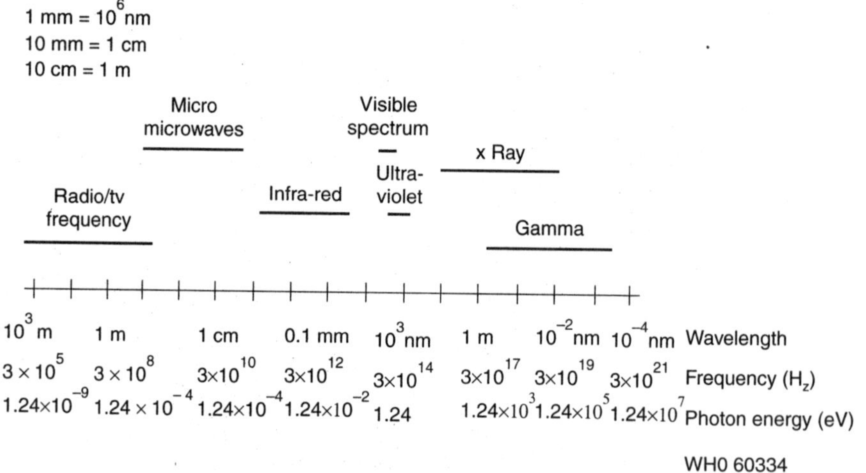

Fig 3. 2 The spectrum of electromagnetic radiation – a WHO document

Conventionally a photon energy of 12eV, corresponding to a wavelength of 100 nm, is taken as the dividing line between ionizing and non-ionizing radiation. This is in the vacuum region of the ultraviolet spectrum. Microwave and RF radiations are much less energetic. Their energy per photon corresponds to 1.25×10^{-3} eV at 300 GHz and 4.1×10^{-10} eV at 100 kHz, and is much too low to cause ionization.

3.1.1 Units of Radiation

When microwave or RF radiation is absorbed in a medium, the most obvious effect is heating. The radiation intensity can be determined calorimetrically. In SI terminology, it is known as the irradiance and is expressed in W/m^2. Traditionally, however, the term "power density" has been and continues to be used for this part of the frequency range, and will be used in this chapter with the more commonly reported units of mW/cm^2 and $\mu W/cm^2$.

The associated electric and magnetic field strengths *(E* and *H)* can be equally valid expressions of radiant energy flow. When these are stated in V/m and A/m, respectively, their product yields VA/m^2. At distances greater than about one wavelength from the source, *E* and *H* are in phase and VA/m^2 may be expressed as W/m^2. Ideally, at a distance sufficiently remote from the source of radiation that it can be regarded as a point source, an inverse square law of power density with distance applies; the ratio E/H is 120 pi, i.e., 377OMEGA. The power density can, therefore, be derived from $E^2/377$ or from $H^2 \times 377$. Where *E* and *H* are expressed in V/m and A/m (Table 3.1), this is referred to as plane-wave or far-field conditions and to obtain a measure of the radiated power density, only the *E* field or the *H* field need be measured. Most instruments used for measuring power density measure the *E* field, because this technique is more versatile and presents fewer practical problems. *H* field detectors have been devised for a limited range of frequencies. Instruments combining both types of detection are possible, in principle, but would be most difficult to construct.

Table 3.1 Comparison of power densities in the more commonly used units for free-space

W/m^2	mW/cm^2	$\mu W/cm^2$	V/m	A/m
10^{-2}	10^{-3}	1	2	5×10^{-3}
10^{-1}	10^{-2}	10	6	5×10^{-2}
1	10^{-1}	10^2	2×10	5×10^{-2}
10	1	10^3	6×10	1.5×10^{-1}
10^2	10	10^4	2×10^2	5×10^{-1}
10^3	10^2	10^5	6×10^2	1.5
10^4	10^3	10^6	2×10^2	5

The distance beyond which far-field conditions apply is usually taken as being $2a^2$/lambda, where a is the maximum dimension of the source (antenna) and lambda is the wavelength. The radiated "near field" includes distances of less than $2a^2$/lambda, where the inverse square law with distance does not apply and the impedance in space (the ratio E/H) may differ from 377OMEGA. Close to the source, at distances less than lambda, reactive components of E and H become progressively more important. Instruments, calibrated in units of power density but based on the measurement of E, for instance, will become increasingly inaccurate at close range. The instruments make valid measurements of the E fields, but their scale indications in terms of power density no longer apply.

3.2 ELECTRO MAGNETIC INTERFERENCE AND RADIATION AS CAUSE OF CASUALTY

During the last few decades, the use of electric and electronic systems for communication and other applications is increased by many folds. In compact environments like boats, ships, aircrafts, it is necessary for different devices to operate in close proximity of each other. Such devices often affect performance of other near by devices adversely. This happens with the coupling of their signals through near and far regions, propagating electromagnetic fields. Thus EMI has become a major problem in most of the situations. The presence of considerable amount of EMI makes the susceptible systems to become inoperable resulting in the collision of sea vessels and other accidents.

3.2.1 General Effects of EMI and EMR

Electromagnetic interference is a form of environmental pollution. Specialized circuits are becoming increasingly vulnerable to pulsed electromagnetic interference, such as ESD, EFT, SURGE, and power quality failure transients. These are both visible and invisible, and range in intensity from lightning bolts to normal switch arcing, to small but potentially disastrous electrostatic discharge from the human touch or even furniture. The sparks involve very high power or extremely fast transients, thus effecting quality of ac power mains, [DNV Standards].

In sea-going vessels, this frequency spectrum pollution emanate out of

1. Electrostatic discharges in nature
2. High voltage cabling
3. Radar's
4. Electric motors, actuators
5. Ignition system of engine
6. Wireless and communication systems
7. Marine communication antennas
8. House keeping appliances, cooking appliances
9. Lamps of special types
10. L.V. fuses and H.V. fuses
11. Static watt-hour meters

In view of such devastating effects in small and big vessels, it is of interest in this chapter to present a brief account of EMR and its effects as well as the steps taken for managing possible EMI which would be inputs into a safety index given later in chapter-5. The safety index indicates a level of survivability of a vessel and its crew in a hostile environment affected by design deficiencies, human and organizational errors, weather and effects of EMI.

The main source of EMI in small and big vessels is from the intentional radiators like communication and radar antennas and the other power electrical and electronic equipment that are packed in the communication room and engine room of vessels. Switches, whether they are electrical, thermostatic, electromechanical or electronic, present in such environments, create interference with their ability to change their impedance from zero to infinity or from infinity to zero. When large currents are involved, the resultant EMI is large. Similarly EMI produced from electronic/electrical equipment such as discharge tubes, rectifiers, transformers, mercury vapor lamps, sensors, diodes etc. , affects the communications and other systems. The narrow band or broadband EMI produced from different sources affects the susceptible devices either by radiation or by conduction. In fact, EMI produced from natural sources like cosmic and atmospheric electrical disturbances are not controlled and it is necessary to protect the susceptible devices using one or more of the following EMI design Techniques; shielding, filtering, isolation, separation and orientation, circuit impedance level control, cable design and cancellation techniques (frequency or time domain). On the other hand, the man made EMI can either be controlled or eliminated. In situations where it is not possible to eliminate completely, the susceptible devices must be protected using EMC techniques.

An important source of interference in ships is ground loop. It is usually a common connection of electrical circuits to a conducting medium that becomes a common reference plane. Current supplied to circuits from the power supply do not disappear in the equipment. They must return to the source through the chassis or through ground leads. It is important that these currents return by individual low impedance paths in order to reduce coupling between high and low level signals. With respect to grounding, the radio frequency portion of the spectrum is the most difficult and complex to work and complexity varies in proportion to operating frequency.

In addition to the EMI discussed above, a vessel has external sources of EMI. Big and small vessels operate with its own problems of electromagnetic environment in an all enclosed, conducting steel sitting in one of the finest conducting mediums in the world. i.e. water. With a vessel's long run of electric cables and its many elaborate antennas as well as many unexpected antennas such as ladders, railings, and projections from the hull structure, it is a very poor example of electromagnetic compatible design. Most vessels operate close to each other in channels and can strongly interfere with each other's signals.

Natural interference phenomena though erratic, do exist. This interference occurs predominately in frequency ranges below 50 MHz. It dominates all other interference sources below 30MHz and is usually the limiting factor in communications and propagation of the incident energy. Another property of EMR results in the induction or coupling of energy into objects such as vessels rigging, fences, and other antennas. These produce involuntary and uncontrolled reaction or ignite clothing worn by the victims. The major damage to humans is from thermal effects which are pronounced in the region of 100 to 3000 MHz and are a function of intensity and frequency of incident energy.

The penetration of energy into the body and its absorption's and reflections will depend not only upon the physical dimensions and dielectric constant of the tissues, but also upon the EM radiation[British Standards].

The temperature-Humidity index (THI) is a factor sometimes used to modify the safe level of exposure to man when internal heating could produce prostration for a high THI environment. Body fat may act as quarter wave-matching transformers to match the impedance of air to the input impedance of deep tissues. It has been noted that sedatives and tranquilizers interfere with the body's ability to regulate temperature and lose heat, thus lowering the threshold of tolerance.

A complete resume of biological effects of microwave damaging levels are given in Table 3.2.

Table 3.2 Resume of Biological effects of microwave at damaging levels.

Frequency MHz	*Wavelength (cm)*	*Site of major tissue effects*	*Major biological effects*
100	Above 200	Not established-probably whole body	General warming of exposed areas (used diathermy)
150-1,200	200-25	Internal body organs	Damage to internal organs from overheating
1,000-3,300	30-10	Lens of the eye	Lens of the eye susceptible and tissue heating
3,300-10,000	10-3	Top layers of the skin, lens of the eye	Skin heating with the sensation of warmth
10-100 GHz	Less than 3	Skin	Skin surface acts as reflector or absorber with heating effects

* Damaging levels vary with frequency, ambient temperatures, and individuals. Safety criteria establish levels above 10 mW/cm^2 at any frequency as being unsafe – [WHO]

Certain non-thermal effects of EMR have also been noticed in recent times. Some of the recorded non-thermal effects are

1. Minor changes in human blood properties upon exposure to electromagnetic energy of proper frequency and intensity causing temporary instability.
2. Certain people can hear a buzz exposed to microwave radiation, effecting decision making in closed conditions.
3. Abnormalities of the chromosome structure occurring upon exposure have been noticed causing temporary instability.
4. Movement, orientation and polarization of protein molecules in pulsed RF fields have been noted causing temporary instability.
5. Unexplained response of man to radar-Epi-gastric distress, emotional upsets, and nausea have occurred at as low as 5-10 mw/cm^2 and are most commonly associated with the frequency range from 8×10^3 to 12×10^3 MHz.

Aboard vessels, voltages may be induced in standing rigging, rails, and cables, parts, of the superstructure and deck loads. Voltages of sufficient amplitude to cause serious burns to personnel can be induced into handling equipment by induction from HF transmitting antennas. In addition the involuntary reaction of personnel to non-lethal shock is extremely dangerous when a person is working on close quarters or in elevated location since such reflex actions can result in falls or body injury due to striking an object.

The radiation limits that should be adhered to for a frequency range between 100 MHz and 100GHz for non-injury to personnel are

1. 10 mw/cm^2 average incident power energy for exposures greater than 30 seconds.
2. 300 mj/cm^2 for intermitted exposures between 3 and 30 seconds.

In fact, it is possible to analyze and forecast some of the hazards. Two time factors are of primary interest in identifying potentially hazardous space in a complex radiation field.

1. The most common factor is related to moving radiators such as search radar antennas or electronic scanning arrays.
2. The other is for scheduled radiators.

Only the first one is of importance to vessels as far as radiated energy from antennas is concerned and the second variety is present in the engine room, cold packaging room, communication room, etc.

3.2.2 Analysis of Electromagnetic Radiation on Sea-Going Vessels

The first step in finding out the energy levels of EMR is to do an analysis by

1. First theoretical optical modes of designating the suspected field energies.
2. Field measurements.

In theoretical prediction complex energy levels owing to a number of devices may be combined either by way of power derivatives or field strength. Any drastic difference, between measured and predicted radiation levels should not be disregarded since the inaccuracies involved in the prediction process are present. Whenever there is a reason for doubt or conflict, measured value should be given precedence in the decision making process. The combination of measured and predicted results should remain as real factor in the final determination of existing hazards.

To predict the possible effects of EMR it is necessary to know the EM environment in vessels. Typical values of EMI on board a vessel is given below [EM - 50081Standards].

Table 3.3 Typical EMI values on board vessels

Location	*Value*
Wheel house top	5 to 80 V/m at 10 cm above the steel deck 120 to > 200 V/m at 2m above the steel deck 4 V/m at the window 0.5 to 1 V/m at the center console
Bridge wings	3 to13 V/m at 10 cm above the steel deck 50 to 200 V/m at 2 m above the steel deck
Open deck Under the	0 to 30 V/m 480 KHz
Antenna on fore peak	0 to 7 V/m; 4,18 MHz 200 V/m at 2m above the steel deck
Accommodation	.0.1 V/m near handrail on stairs
Machinery	0.1 V/m

Here EMI covers the spectrum from about 10 Hz to 100 GHz. For radiation emission a lower frequency limit of 10 KHz is often used. There is no pure DC EMI except from Electrostatic discharge. In favorable conditions the static discharge can approach 25 KV in magnitude, but normally not more than 6 KV by contact and 8 KV in air.

3.3 ELECTROMAGNETIC RADIATION HAZARD

EMR can be sensed by humans in many instances. Research has shown that radiation can make serious changes in human's biochemistry. The effects are numerous.

Electromagnetic radiation has potential for hazards and effects of various forms; EMR has biological effects on human tissues and organs and the properties and condition under which it can produce adverse effects. Also EMR is potentially hazardous to various volatile fuels such as the HSDS and HFSDS used in vessel, gas used for cooking.

3.3.1 Biological Hazard Effects

Radio waves, light, heat, x-ray, and gamma rays are all forms of electromagnetic radiation differing only in their wavelength and energy. In order to produce harmful effects electromagnetic radiation must cause physical or chemical changes in tissue or displace atoms in solid crystal type materials thereby destroying effectiveness of the device. For EMR to interact with tissue, factors that are crucial are intensity, frequency, energy level polarization and duration of exposure. Response will be either thermal heating of organs or non-thermal. Once EMR radiation enters the tissue the dielectric and conductive properties of the tissue alters the absorption and propagation of the incident energy. Another property of EMR results in the induction or coupling of energy into objects such as Vessel rigging, fences, and other antennas. These produce involuntary and uncontrolled reaction or ignite clothing worn by the Seafarers. The major damage to humans is from thermal effects which are pronounced in the region of 100 to 3000 MHZ and which are a function of intensity and frequency of incident energy.

The penetration of energy into the body and its absorptions and reflections will depend not only upon the physical dimensions and dielectric constant of the tissue, but also upon the EM radiation. The Temperative – Humidity index (THI) is a factor sometimes used to modify the safe level of exposure to man when internal heating could produce prostration for a high THI environment. Body fat may act as a quarter – wave matching transformer to match impedances to air to the input impendence of deep tissues. It is interesting to note that sedatives and tranquilizers interfere with the body's ability to regulate. Temperature and loss of heat thus lowering the threshold of tolerance. Tabulation below gives the relationship of physical size to wavelength.

Table 3.4 Comparisons of frequency Wavelength and Equivalent Number of Wavelengths of Man 1.7 meters Tall (5' 7")

Frequency (MHz)	*Wavelength (Meters)*	*(cm)*	*Equivalent Number of Wavelengths*
3	100	10000	0.017
30	10	1000	0.17
300	1	100	1.7
3000	0.1	10	17.0
10,000	0.03	3	56.6

3.3.2 Other Radiation Hazard

EMR radiation of high enough levels can cause physical as well as permanent damage to communications and electronics equipment. The mechanisms leading to damage of the equipment from RF energy are complex. Damage commonly occurs at the circuit component level and may occur from direct exposure to radiation via thermal heating or more probably from voltages or currents induced by electro-magnetic fields at antenna terminals circuit wiring component terminals, power lines etc.

Transistors and other semi conductors including integrated circuits, microelectronics is specially susceptible to damage, from fast transients where peak - induced voltage exceed the minimum rating of the device. Diodes are also subject to reverse breakdown by induced RF Voltage in excess of devised rating.

The EMI coupling mechanisms on board a vessel are as shown.

1. Power cable conducted emission
2. Power cable conducted susceptibility
3. Interconnecting cable conducted emission
4. Interconnecting cable conducted susceptibility
5. Antenna lead conducted emission
6. Antenna lead conducted susceptibility
7. Common ground impedance emission coupling
8. Common ground impedance susceptibility coupling
9. H –field radiation
10. E-field radiation
11. H – field susceptibility
12. E-field susceptibility

The general standards for field intensity and radiated power from antenna in free space of Vessels are given by the following:

1. The Power density S at a part due to the power PT radiated by an isotropic radiator

$$S = P_T/4\pi r^2 \quad \text{and} \quad S = E^2/120\pi$$

Where S = Power density [W/m^2]

r = distance $[m]$

P_r = Transmitted power $[W]$

The Field intensity of an isotropic radiation in free space is

$$E = 5.5\ (P_T)^{1/2}/r$$

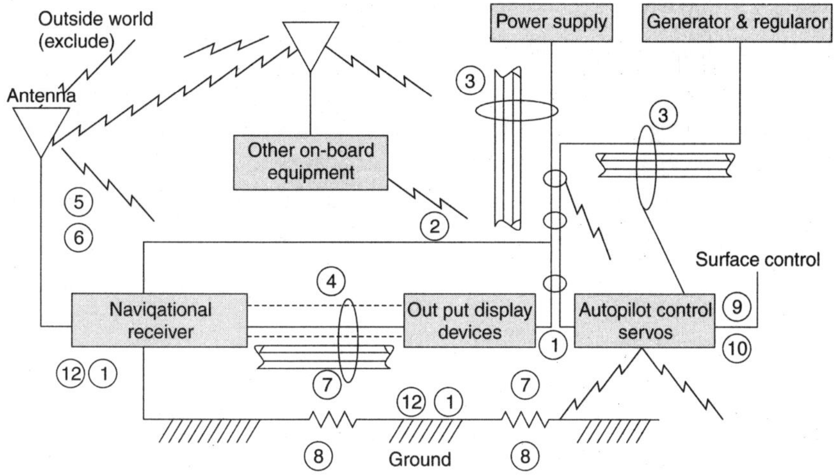

Fig. 3.3 EMI coupling mechanism on board a vessel

Fig. 3.4 Typical noise path

2. For a half wave dipole in the direction of maximum radiation

$$S = 64P_T/4\pi r^2, E = 7.01(P_T)^{1/2}/r$$

Where 1.64 = maximum gain of a half wave dipole

3. Shielding effectiveness

$$SE = A + R + B$$

Where A = Absorption loss

R = Reflection loss

B = Secondary loss

3.4 SHIELDING EFFECTIVENESS OF MAGNETIC AND NON-MAGNETIC MATERIALS

The effect of EMI can be reduced if it is controlled from design stage, capacitors, inductors, electro mechanical devices, ferrite components, EMI gaskets cabling and connectors must meet specifications, though 100% reduction is theoretically impossible. The installation factor is more critical and this is where human error creeps in.

Table 3.5 Shielding effectiveness of some magnetic and non-magnetic materials

Material	*Freuqency*	*Absorption*	*Reflection loss*		
	(Khz)	*Loss[(1)] all fields*	*Magnetic[(2)] field*	*Electric field*	*Plane wave*
Magnetic	< 1	0-30dB	0-10dB	> 90dB	> 90dB
μ_r = 1000	1-10	30-90dB	0-30dB	> 90dB	> 90dB
	10-100	> 90dB	10-30dB	> 90dB	60-90dB
	> 100	> 90dB	10-60dB	60-90dB	30-90dB
Non magnetic	> 1	0-10dB	10-30dB	> 90dB	> 90dB
μ_r = 1	1-10	0-10dB	30-60dB	> 90dB	> 90dB
	10-100	10-30dB	30-60dB	> 90dB	> 90dB
	> 100	30-90dB	60-90dB	> 90dB	> 9dB

1. Absorption loss for 0, 8 mm thick shield.
2. Magnetic field reflection loss for a source distance of 1m). Explanation: 0-10dB = Bad; 10-30dB = poor 30-60dB = average; 60-90dB = Good; > 90dB = Excellent

3.5 MAGNITUDE OF EXPOSURE TO MICROWAVE AND RF RADIATION AND SOURCES OF CONCERN

Electromagnetic fields and RF radiation occur naturally over a very wide range of frequencies. The ionosphere very effectively shields the earth's biosphere from radiations of this type originating in space. Electromagnetic fields and radiation of high intensity may be generated by natural electrical phenomena such as those accompanying thunderstorms.

However, in the frequency range of 100 kHz to 300 GHz, the intensity of natural fields and radiation is low. Exposure of the urban population to man-made microwave sources was found to vary from a very low value to as high as 100 $\mu W/cm^2$. The median exposure to the total microwave flux from external sources for a population was calculated to be 0.005 $\mu W/cm^2$. It has been observed that the calculated background exposure from the sun, integrated up to 300 GHz would be in the range of 1.4×10^{-5} $\mu W/cm^2$. These values can be put in better perspective by noting that the integrated microwave/RF flux emitted from the human body has been calculated to be up to 0.5 $\mu W/cm^2$.

The proliferation of man-made sources of energy in the 100 kHz-300 GHz range has only occurred over the last few decades. From the point of view of biological evolution, this energy constitutes a very recent physical factor in the environment. Observations of biological effects from exposure to microwaves gave rise to concern in the early 1940s. On the basis of special research programmes, radiation protection guides recommending exposure limits were developed in the 1950s in the USSR and the USA and recently in India. Thereafter, several industrialized countries introduced recommendations and/or legislation on microwave and RF health protection.

It should be noted, however, that exposure limits vary widely, and are the subject of many discussions and much controversy.

Although concern about microwave and RF effects and possible hazards arose first in highly developed countries, the problem is universal. Developing countries are rapidly establishing telecommunications, broadcasting systems, and other sources of electromagnetic energy. Electromagnetic waves emitted in particular countries may propagate around the globe. A report from the ITU states: "Unless adequate monitoring programs and methods of control are instituted in the near future, man may soon enter an era of energy pollution comparable to that of chemical pollution of today."

The urgent need for international agreement on maximum exposure limits and international programmes for the containment of electromagnetic pollution has been stressed at international meetings. Prevention of potential hazards is a more efficient and economical way of achieving control than belated efforts to reduce existing levels.

3.6 ENVIRONMENTAL HEALTH CRITERIA FOR RADIOFREQUENCY AND MICROWAVES

In November 1971, the WHO Regional Office for Europe convened a Working Group meeting in The Hague which recommended, *inter alia,* that the protection of man from microwave radiation hazards should be considered a priority activity in the field of non-ionizing radiation protection. To implement these recommendations, the Regional Office decided to prepare the manual on "Non-Ionizing Radiation Protection", which will include a chapter on microwave radiation (WHO, 1981).

A Joint WHO/IRPA Task Group on Environmental Health Criteria for Radio frequency and Microwaves met in Geneva from 18-22 December 1978. The Task Group reviewed and revised the draft criteria document, made an evaluation of the health risks of exposure to radio frequency and microwaves, and considered rationales for the development of exposure limits.

In 1973, a symposium, sponsored by WHO, was held in Warsaw, on the biological effects and health hazards of microwave radiation. This symposium provided one of the first opportunities for an international exchange of quite diverse opinions on the effects of microwaves. Recommendations adopted by the symposium included the promotion and coordination, at an international level, of research on the biological effects of microwaves, and the development of a non-ionizing programme by an international agency.

The International Radiation Protection Association (IRPA) became responsible for these activities by forming a Working Group on Non-Ionizing

Radiation at its meeting in Washington, DC, in 1974. This Working Group later became the International Non-Ionizing Radiation Committee (IRPA/INIRC) at the IRPA meeting in Paris in 1977.

Two WHO Collaborating Centres, the National Research Institute of Mother and Child, Warsaw (for Biological Effects of Non-Ionizing Radiation) and the Bureau of Radiological Health, Rockville (for the Standardization of Non-Ionizing Radiation) cooperated with the IRPA/INIRC in initiating the preparation of the criteria document. The final draft was prepared as a result of several working group meetings, taking into account comments received from the national focal points for the WHO Environmental Health Criteria Programme in Australia, Canada, Japan, Netherlands, New Zealand, Poland, Sweden, the United Kingdom, and the USA as well as from the United Nations Environment Programme, the United Nations Industrial Development Organization, the International Labour Organization, the Food and Agriculture Organization of the United Nations, the United Nations Educational, Scientific and Cultural Organization, and the International Atomic Energy Agency. Modern advances in science and technology change man's environment, introducing new factors which besides their intended beneficial uses may also have untoward side effects. Both the general public and health authorities are aware of the dangers of pollution by chemicals, ionizing radiation, and noise, and of the need to take appropriate steps for effective control. The increasing use of electrical and electronic devices, including the rapid growth of telecommunication systems (e.g., satellite systems), radio broadcasting, television transmitters, and radar installations has increased the possibility of human exposure to electromagnetic energy and at the same time, concern about possible health effects.

This chapter provides some information on the physical aspects of electromagnetic fields and radio waves in the frequency range of 100 kHz-300 GHz, which has been arbitrarily subdivided according to the traditional approach into microwaves (300 MHz to 300 GHz) and radio frequencies (100 kHz to 300 MHz). It is known that electromagnetic energy in this frequency range interacts with biological systems and a summary of knowledge on biological effects and health aspects has been included. In a few countries, concern about occupational and public health aspects has led to the development of radiation protection guides and the establishment of exposure limits. Several countries are considering the introduction of recommendations or legislation concerned with protection against the untoward effects of non-ionizing energy in this frequency range. In others, the tendency is to revise existing standards and to adopt less divergent exposure limits.

3.6.1 Physical Characteristics in Relation to Biological Effects

Microwave and radio frequency (RF) radiation constitute part of the whole electromagnetic spectrum. The area of concern is with frequencies lying between 10^5 and 3×10^{11} Hz (100 kHz and 300 GHz). The term radio frequency refers to the range 100 kHz-300 MHz (3 km to 1 m wavelength in air) and microwaves to the frequency range of 300 MHz-300 GHz (1 m to 1 mm wavelength in air).

Exposure conditions in the microwave range are usually described in terms of "power density" and are reported in most studies in watts per square meter, or mill watts or microwatts per square centimeter (W/m^2, mW/cm^2, $\mu W/cm^2$). However, close to microwave and RF sources with longer wavelengths, the values of both the electric (V/m) and magnetic (A/m) field strengths provide a more appropriate description of the radiation. Exposure conditions can be altered considerably by the presence of objects, the degree of perturbation depending on their size, shape, orientation in the field, and electrical properties. Very complex field distributions can occur, both inside and outside biological systems exposed to microwaves and RF. Refraction of the radiation within these systems can focus the transmitted radiation resulting in markedly no uniform fields and energy deposition. Different energy absorption rates may result in thermal gradients causing biological effects that may be generated locally, difficult to predict, and perhaps unique.

When electromagnetic radiation passes from one medium to another, it can be reflected, refracted, transmitted, or absorbed, depending on the biological system and the frequency of the radiation. Absorbed microwave and RF energy can be converted to other forms of energy and cause interference with the functioning of the living system. Most of this energy is converted into heat. However, not all microwave and RF radiation effects can be explained in terms of the biophysical mechanisms of energy absorption and conversion to heat. It has been demonstrated both theoretically and experimentally that other types of energy conversion are possible. Interactions at the microscopic level leading to perturbations in complex macromolecular biological systems (cell membranes, sub cellular structures) have been postulated. Biological phenomena caused by such perturbations are expected to show resonant frequency dependence.

3.6.2 Sources and Control of Exposure

General population exposure from man-made sources of microwave and RF radiation now exceeds that from natural sources by many orders of magnitude. The rapid proliferation of such sources and the substantial increase in radiation levels is likely to produce "electromagnetic pollution".

Major man-made sources include: radar installations, broadcasting and television networks, and telecommunication equipment. In industrial, commercial, and home equipment, notably those where energy is applied for heating purposes such as plastic sealing, welding, drying, cooking, and defrosting, there may be extraneous emission (leakage) of microwave or RF radiation.

The problem of this extraneous radiation or pollution from sources of 100 kHz-300 GHz electromagnetic waves varies from country to country, depending on the degree of industrialization. Radiation emitted from high power sources such as broadcasting and telecommunication networks propagates over large distances and may even cover the whole circumference of the globe. With the increasing use of transmitter/receivers by sea and air traffic, and the necessity for ground-based radar control, increased levels of environmental electromagnetic radiation may constitute a problem in many countries.

Problems of pollution range from electromagnetic interference, particularly in relation to the operation of health services, to direct risks to the health of individuals, specially for seafarers who are confined to potentially hazardous emission spaces for a long time.

3.6.3 Power Density Ranges in Relation to Health Effects

During International Symposiums on biological effects and health hazards of microwave radiation, it was agreed that microwave power densities could be divided into ranges. The following is an abridged version of this agreement:

Microwave densities may be divided into the following 3 ranges:

(a) High power densities, generally greater than 10 mW/cm^2, at which distinct thermal effects predominate;

(b) Medium power densities, between 1-10 mW/cm^2, where weak but noticeable thermal effects exist; and

(c) Low power densities, below 1 mW/cm^2, where thermal effects are improbable, or at least do not predominate.

The boundaries indicated for these ranges are arbitrary and depend on numerous factors, such as human size, threshold of warmth sensation, frequency, and pulsing. The introduction of the intermediate range of subtle effects calls attention to the need for additional research, aimed at clarification of the underlying mechanisms.

It should be noted that the classification is applied to the microwave region (300 MHz-300 GHz). A similar classification was not determined for the RF region (100 KHz-300 MHz) because of the nature of its effects.

3.6.4 Exposure Effects in Man

The meager evidence available on exposure effects in man has been obtained from incidents of accidental acute over-exposure to microwaves and RF. Not enough attention has been given to the conduct of epidemiological investigations. In some human studies, which have been conducted on people exposed occupationally, subjective symptoms have been reported.

A considerable number of people in many countries have received microwave and RF diathermy treatment at power levels of several tens of watts for duration of about 20 min daily over a period of some weeks. During physiotherapy or other radiological based treatments adverse effects have not been adequately investigated among diathermy patients. This is a group of people exposed to microwaves and RF who can be readily identified and such studies should be carried out, as they may yield considerable information concerning exposure effects in man.

3.6.5 Health Risk Evaluation as a Basis for Exposure Limits

Theoretical considerations, experimental animal studies, and limited human occupational exposure data constitute the basis for the establishment of health protection standards. It should be noted that, in some countries, microwave and RF health protection standards have recently been changed and that there is a tendency to adopt less divergent exposure limits in comparison with those proposed two decades ago.

In establishing health protection standards, different approaches and philosophies have been adopted.

A highly conservative approach would be to keep exposure limits close to natural background levels. However, this is not technically feasible. A reasonable risk-benefit analysis has to be considered. In ships for example the risk benefit analysis must be clearly put forth and the seafarer compensated for risky ventures.

More data on the relationship between biological and health effects and the frequency and mode of generation of the radiation, particularly in complex modulations, are needed.

Post service medical checkups of seafarers must be initiated to reveal any such exposure.

In the case of pulse modulation, peak power density may be a factor which should be considered in setting exposure limits. However, it is not possible to propose a limit of peak power density from the information available at present.

3.7 RECOMMENDATIONS FOR FURTHER STUDIES, EXPOSURE LIMITS, AND PROTECTIVE MEASURES

3.7.1 General Recommendations

(a) The basic biophysical mechanisms of interaction of microwaves and RF with living systems still need clarification and further studies.

(b) Work on both theoretical and experimental dosimetry, the calculation and measurement of fields and of energy deposited within simulated or actual biological systems, should be continued and refined.

(c) Results of animal studies are difficult to extrapolate to man, and these studies alone do not constitute a satisfactory basis for the establishment of health protection criteria. They should, therefore, be supplemented by appropriate epidemiological studies in man.

(d) The existing data on power, amplitude, and frequency "windows" seem to warrant continued investigations.

(e) The effects of chronic exposure on sensitivity to convulsant and other drugs are potentially useful and may have a direct bearing on the development of exposure standards.

(f) Long-term, low-level exposures combined with such stresses as high ambient temperature and humidity such as typical on ships should be investigated.

(g) There is little published information on dose-effect relationships; reports tend to be limited to whether effects are observed at one particular level of exposure rather than over a range.

(h) More dose-related information, even covering small subject areas, would be valuable.

(i) Investigations on the genetic effects and effects on development of microwave and RF radiation should have priority.

(j) Attention should be given to investigating the different sensitivities to microwave/RF exposure of subgroups within the general population.

(k) National and international agreements on exposure limits, ways and means of controlling this type of environmental pollution, and concerted efforts to implement such agreements are needed.

3.7.2 Measurement Techniques

(a) There is a continuing need for the development of microwave/RF measuring instruments that: (a) give direct readings of electric or magnetic field strength, or power density; (b) are robust; (c) are portable, light-weight, and battery-operated; and (d) are sensitive and can be used over a wide frequency range.

(b) The problem of the design of personal dosimeters which could be useful on ships also remains to be solved.

3.7.3 Safety Procedures

(a) Computation techniques or methods that predict the distribution of fields close to deliberate high-power emitters (in the near field) are needed.

(b) Emphasis should be placed on the development of technology to ensure containment and limitation of radiation to the deliberately exposed object, as well as the reduction of leakage emission from devices.

(c) Personal protective devices should be used only as a last resort.

(d) Adequate medical surveillance of occupationally exposed persons should be provided, specially in form of post sea medical tests

(e) Once exposure limits have been set, safety guidelines or codes of practice concerning safe use and installation design should be developed as soon as possible.

3.7.4 Biological Investigations

Reports of experimental work should contain sufficient information describing the exposure conditions to allow an estimation not only of the total absorbed energy but also, as far as possible, of the distribution of the energy deposited within the irradiated biological system.

Systematic investigations of the effects of microwave/RF exposure at all levels of biological organization are to be encouraged. This includes effects at the molecular level on sub cellular components; cells, viruses, and bacteria; organs and tissues; and whole animals.

Particular attention should be paid to: (a) long-term, low-level exposures and possible delayed effects; (b) the possibility of differences in sensitivity of various body organs and systems, where specific effects in various animal species are being considered; and (c) the influence of microwave/RF exposure on the course of various diseases, including any possible increase in sensitivity to microwaves/RF that may result because of the disease state.

3.7.5 Epidemiological Investigations

(a) Epidemiological studies should be carried out in a careful manner, paying attention to the relationship between exposure to microwaves/RF and other environmental factors occurring in the place of work and to the health status of the investigated group. Specific biological endpoints should be selected and adequate examination methods used for such studies. Conventional medical examinations will not provide sufficient information.

(b) Studies should be carried out on (a) seafarers occupationally exposed to microwave/RF sources; (b) patients treated with microwave and RF diathermy; and (c) groups within the general population living near high-power microwave/RF sources, such as high tension lines, antennas and Radars.

(c) A distinction should be made between occupational and public health protection standards.

3.8 EXPOSURE LIMITS AND EMISSION STANDARDS

3.8.1 Occupational Exposure Limits

The occupationally exposed population like seafarers consists of healthy adults exposed under uncontrolled and controlled conditions, and who are unaware of the occupational risk. The exposure of this population should be monitored.

It is possible to indicate exposure limits from available information on biological effects, health effects, and risk evaluation. For occupational exposure, values within the range 0.1-1 mW/cm^2 include a high enough safety factor to allow continuous exposure to any part of the frequency range over the whole working day. Higher exposures may be permissible over part of the frequency range and for intermittent or occasional exposures.

3.8.2 Exposure Limits for Seafarers

The general population on ships includes persons of different age groups and different states of health, including women and children. The possibility that these people could be particularly susceptible to microwave/RF exposure deserves special consideration.

Exposure of the ships crew should be kept as low as readily achievable and exposure limits should generally be lower than those for occupational exposure.

3.8.3 Emission Standards

Emission standards for equipment should be derived from, and be lower than exposure limits, where this can reasonably be achieved. A class of equipment may be considered safe and exempt from regulations, if hazardous levels of radiation exposure cannot originate from such source, that is preferably with a statutory certification label, certifying the standards of tests.

3.8.4 Implementation of Standards

The implementation of microwave and RF occupational and public health protection standards necessitates: the allocation of responsibility for measurements of radiation intensity and interpretation of results; and the establishment of detailed radiation protection safety codes and guides for safe use, which indicate, where appropriate, ways and means of reducing exposure.

3.8.5 Other Protective Measures

Prevention of health hazards related to microwave and RF radiation also necessitates the establishment of rules for the prevention of interference with medical electronic equipment and devices such as cardiac pacemakers, prevention of detonation of electro explosive devices, and prevention of fires and explosions due to the ignition of flammable material (vapours) by sparks originating from induced fields. This is a special area of concern on tankers, gas carriers, chemical carriers, and fueling operations on board any ship. Use of mobiles VHF equipment near any volatile or gaseous environment including closed environments as in ship tanks can cause catastrophic results.

3.8.6 Studies Related to the Establishment of Limits

Studies of the frequency and modulation dependence of biological and health effects are of prime importance. The results of such investigations may make it possible to modify the rationales of present day standards and to identify frequencies at which exposure limits should be lower or higher.

3.9 EMP

Since the early 1980's, the developed countries have been funding scientific and technological programs that seek to develop radio frequency and high power microwave technologies as a potential class of directed energy weapon systems. These technologies are commonly grouped under the "umbrella" of high power microwave (HPM) technologies. The electromagnetic frequency spectrum for this area ranges from the low megahertz to the high gigahertz frequencies ($1 \text{ X } 10^6$ hertz to $1 \text{ x } 10^{11}$ hertz). Invisible to the human eye, these frequencies range from wavelengths of 0.1 centimeters (the gigahertz frequencies) to 3 meters (the megahertz frequencies) in length.

It is not surprising that all the major powers are interested in high power microwave technology. This technology is an outgrowth of previous military and civilian research and studies in the field of radar technology and electromagnetic pulse (EMP) that started in the late 1930's and continued through the late 1980's. EMP encompasses frequencies between the low hertz to high megahertz frequency range. While it is typically equated with nuclear detonations, EMP is also produced by non-nuclear sources. One example is the static and distorted radio signals that occur when a car is driven beneath high voltage power lines. While this effect only disrupts the signals and does not harm the radio, in fact, EMP can produce such serious and sometimes catastrophic effects in various electronics equipment that the Department of Defense developed various hardening and shielding efforts to protect its weapons systems and subsystems against the effects of EMP. EMP is potentially devastating both in its effects on equipment and humans

Chapter 4

EMC Testing
(for validation of marine equipment)

4.1 EMC DISCIPLINES

Why is understanding all the EMC discipline important? The typical EMC problems can involve any combination of the disciplines, as well as combinations of frequencies, dimensions of components, wiring harnesses, and assemblies. This can make problem solving a challenge, requiring seemingly contradictory approaches for solutions at different frequencies or under different conditions.

EMC is divided into three disciplines based on coupling mechanisms. The first is the radiated path, the second is the conducted path, and the third is a combination of the two mechanisms (sometimes encountered with electrostatic discharge [ESD]). Within each area are two additional disciplines-the first emissions, and the second immunity. These are shown in Fig. 4.1.

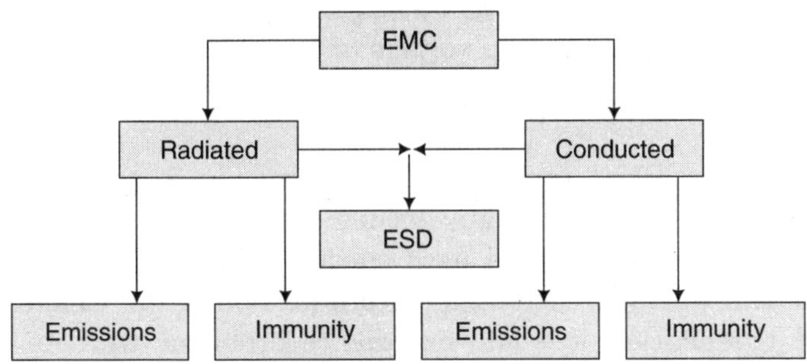

Fig. 4.1 EMC discipline

On the left side of the figure, radiated phenomena are divided into radiated emissions (RE), and radiated immunity (RI).

On the right hand side of the figure, conducted phenomena are also divided into conducted emissions (CE) and immunity (CI).

In the center of the figure is ESD, which may consist of a combination of radiated and conducted phenomena.

Let's examine an example of each one of these disciplines:

If we have a brush-type DC motor, this may affect the reception on a nearby AM radio (you may have heard this in your car). This is an example of "emissions" from the motor and/or its wiring harness, and the "immunity" of the radio not being compatible. Either the emissions of the source must be reduced, or the immunity of the victim must be increased for the radio reception to be free of interference.

The mechanism of coupling in this instance is from the source (motor brushes), radiating via the twelve-volt power wiring harness, which acts as a transmitting antenna. See Fig. 4.2

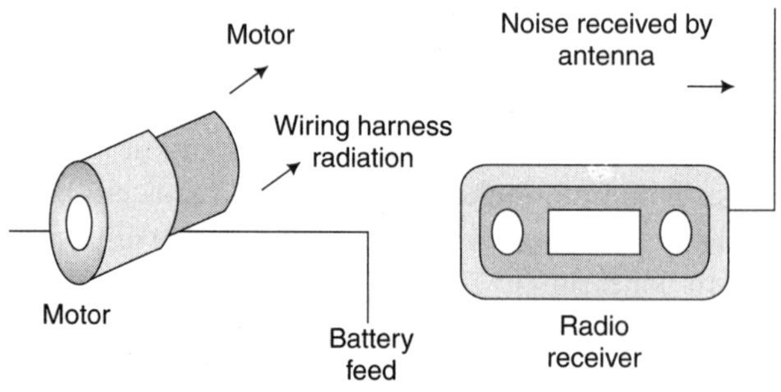

Fig. 4.2 Radiated coupling

Radiated emissions (RE) and radiated immunity (RI) are expressed in terms of microvolt per meter or volts per meter, or their dB equivalent. This means that a 1-meter-long piece of wire going from a 10 V/m point in space to a 20 V/m point would have a voltage of 10 volts across it. For magnetic field coupling, the units would be micro amps per meter or amperes per meter.

How real are EMC problems in shipping as a result of radiated emissions? There have been many documented cases of equipment in ship that was sensitive to radiated path energy. When this particular make of equipments was approaching a certain portion of the Channel, near a broadcast transmitter, it would experience problems with its electronic control system. The solution was to erect a screen around the emitter or around the vessels sensitive equipment to prevent the transmitter energy from being radiated onto the channel. This is an example of a problem with the radiated immunity of the vessel being incompatible with the electromagnetic environment in which the vessel must operate.

Another phenomenon sometimes observed on small vessel radios is "alternator whine". This is an example of "conducted emissions" as shown in Fig. 4.3. The audible ripple frequency voltage travels via the shared power wiring to the power connector on the radio. This is an example of the "conducted emissions" from the vessel, and the "conducted immunity" of the radio not being compatible. Many more such examples exist in the use of oscilloscopes, measuring equipments, scanning equipment, Sensitive probing equipment wired up in an engine room environment. Conducted coupling paths are often much more efficient than radiated coupling paths, which means it takes much more efficient than radiated coupling paths, which means it takes much less energy to cause similar problems.

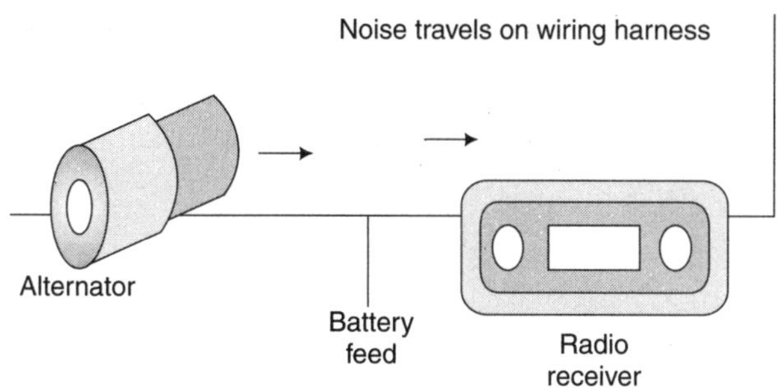

Fig. 4.3 Conducted coupling

Conducted emissions (CE) and conducted immunity (CI) are likewise expressed in terms of volts, or amps, or their dB equivalent.

Now that we've received the basics of EMC and their importance to the shipping industry, we can focus on a particular aspect of EMC disciplines that of the radiated path. In order to discuss the characteristics of the radiated path, it is important to review the fundamentals with regard to electric and magnetic field waves and their propagation. Key to this understanding is called the "right hand rule". If your thumb represents the direction of (conventional) current (positive charges), then as your fingers surround the conductor they will indicate the direction of the magnetic field surrounding the conductor. The electric field radiates outward perpendicular from the conductor, and the magnetic field encircles the conductor. This basic understanding is the key to visualizing the strength and direction of the electric and the magnetic field.

4.2 TEST METHODS

4.2.1 Basic EMC Test Hardware

The objective of EMC testing is to demonstrate conformance to requirements. Failure to meet requirements may indicate a need for redesign; and, typically, further analysis is first performed to determine if the particular failure is likely to cause an EMC problem. For example, an equipment emission that exceeds the radiated emission (RE) limit by 20dB at 100 kHz may not be serious if the overall system for which the equipment is destined does not utilize the spectrum below 2 MHz.

We will discuss some of the basic instrumentation utilized when performing EMC tests on ships equipment. Many of these techniques and instruments are also applicable to EMC in other areas, such as home electronics or military hardware. This section will focus on three main areas, equipment, transmission lines and cables, and EMC measurements.

4.2.2 EMC Instrumentation

The majority of EMC signal generating instrumentation has a source impedance of 50 ohms, as shown in Fig. 4.4.

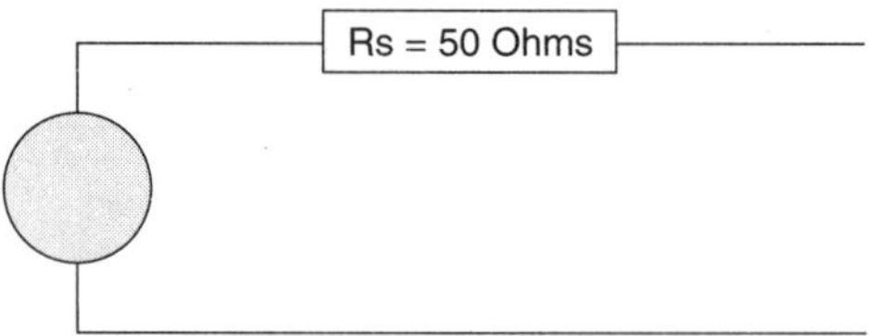

Fig. 4.4 Internal impedance of RF signal source

Most EMC measurement instrumentation has an input impedance of 50 ohms. There are some exceptions to this; volt meters and oscilloscopes may have higher impedance so as to not load down the circuit being monitored during testing.

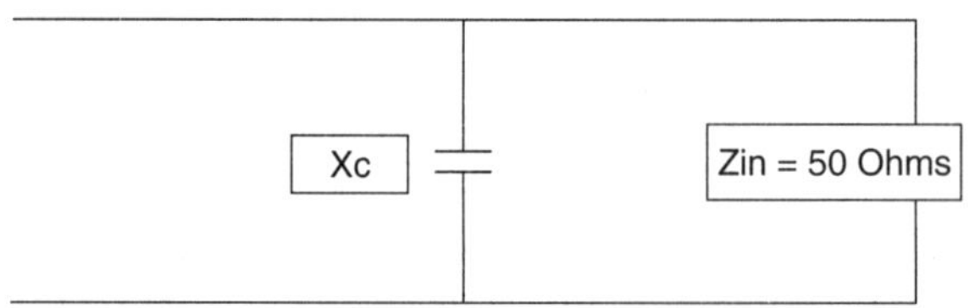

Fig. 4.5 Impedance of most RF measurement instrumentation

Why is 50-ohm impedance so common on test equipment? The reason is that it is intended to match the impedance of the majority of coaxial cables that is used in most ships. There can be complications associated with using impedances other than 50 ohms. The measurement will become frequently dependent, based on this mismatch of cable, source, and load impedances.

4.2.3 Amplifiers

One of the most important components of the RF immunity test system is the power amplifier. Amplifiers are used both for radiated testing using antennas or TEM cells, and for conducted testing using current injection probes or artificial networks. The transducer used has a major effect on the specification of the amplifier, and the two could be considered a system. Amplifiers are often housed in a special room so that adequate cooling can be provided, and so that access can be limited to authorized personnel.

4.2.4 Power Output and Bandwidth

Two specifications, maximum power output and the bandwidth over which this output is sustained, are determined by the required capability of test facility. The bandwidth and power requirements depend on which standards the equipment under test (EUT) will be tested to, and the test configuration.

4.2.5 Frequency Range

European Directive 72/245/EEC covers frequencies from 20 MHz to 1000 MHz. Revisions are being considered to extend radiated immunity testing up to 2 GHz. Many manufacturers test at frequencies outside this range, as well as perform conducted immunity testing, to satisfy their own internal requirements. Any overlap between conducted and radiated testing frequencies depends on internal standards.

The bandwidth may, for instance, be specified as a 3-dB bandwidth, which is likely to mean that only half the rated power is available at the band edges.

4.2.6 Test Level

Assuming that one is working primarily to 72/245/EEC, a test level of 30 V/m with 80 percent modulation is required. Considerations should be given to system losses in cables, connectors, couplers, and switches. Losses increase with frequency but may be offset increasing antenna gain at the higher frequencies.

Table 4.1 Calculated (Watts) for Radiated Testing

	1-Meter Distance			*3-Meter Distance*		
	3 V/m	*10 V/m*	*30 V/m*	*3 V/m*	*10 V/m*	*30 V/m*
27 MHz	25.4	282	254	228.8	2542	22880
80 MHz	1.29	14.3	129	11.59	128	1159
200 MHz	0.29	3.23	29	2.61	29.0	261
1 GHz	0.33	3.64	33	2.95	32.7	295

4.2.7 Conducted Testing

The commonly used bulk current injection (BCI) method requires more power than some other methods since the injection probe is fairly leaky at the extremes of its frequency range. A 100-Watt amplifier is frequently used.

4.2.8 Modulation

Another factor which affects the power requirement is the need for amplitude modulation. This test level is defined in terms of an unmodulated signal; consequently, the field calibration method uses an unmodulated signal. When the test is actually run, modulation is applied. The default modulation is a 1 KHz sinusoid at 80% modulation. The relationship between modulation depth and the amplitude envelop of the signal is shown in Figure 4.6. The peak signal power is increased by 5.2 dB, so the amplitude of the immunity field is reduced by this same amount. This adjusts the peak signal strength with modulation to same peak field strength as without the modulation.

VSWR (voltage standing wave radio) is a measure of the match to a resistive 50 ohms that is presented at the terminals of an amplifier or a transducer. Unless the impedance is exactly 50 ohms, some power is reflected from the terminals and travels back down the feed line.

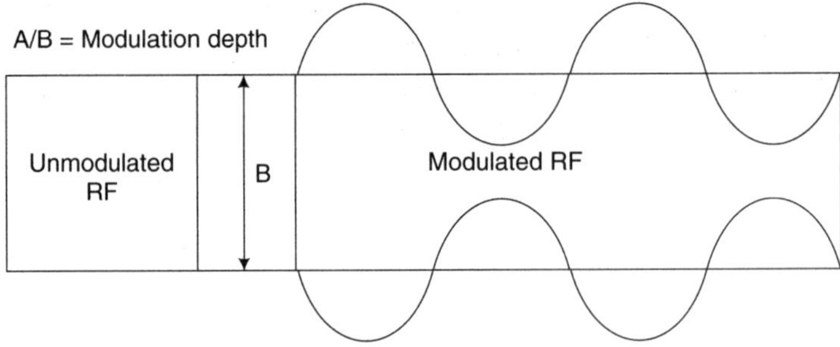

Fig. 4.6 Relationships between modulation depth and amplitude envelope

4.2.9 VSWR Tolerance

A VSWR of 1:1 implies a perfect match. An open or short circuit implies an infinite VSWR. A biconical antenna can show a high VSWR (30:1) at 30 MHz. Even at the higher frequencies, coupling with the screened room and the EUT can increase the VSWRs greater than 60:1 can result. Under these conditions, much of the applied RF power is reflected back to the amplifier output.

A power amplifier for EMC immunity testing must be able to deliver as much of its rated power into mismatched loads as possible.

4.2.10 Class A or AB?

The definitions of Classes A and AB, originally developed for tube-type amplifiers, are not strictly applicable to solid-state amplifiers. An ideal Class A amplifier has an operating current that does not vary with output power. It can be either single-ended or push-pull. In push-pull operation, each valve still conducts through 360^0. The advantage of push-pull is that the non-linearity of the tubes should cancel. Class A has the lowest efficiency, and therefore the highest cost, but, as it is designed to dissipate all the DC power supplied to it when no input signal is present, it can also tolerate total reflection of the output power.

Class A is also the only class of amplifier that will operate with very fast (e.g., pulse) modulation signals. In the other classes the amplifier operating current varies with the modulating signal, and the intrinsic parasitic of the amplifier circuits will not allow rapid variation of the operating current.

Class AB allows for a slight overlap at the crossover point. Each tube conducts over more than 180^0, thus reducing cross cover distortion. A distinction between Classes AB1 and AB2 is the presence or absence of positive grid current. Class AB has better efficiency than Class A, but the amplifier needs to be protected against excess reflected power.

4.2.11 Linearity

When a power amplifier is used for EMC immunity testing with a modulated signal source, the amplifier must remain linear over its entire frequency range up to its maximum power. If it does not, there are two consequences:

The peak of the modulated RF envelope will be flattened. This creates harmonics of the modulation frequency (1 KHz). It will also reduce the actual peak applied field strength.

Harmonics of the carrier frequency will be generated. If the non-linearity is severe, the harmonic amplitudes can approach the fundamental, and this can result in susceptibilities being observed at the wrong frequency. It will also reduce the applied field strength at the carrier frequency. Now combiner technology used in the higher power amplifiers results in exceptionally low harmonic performance, 30 dB or better. IEC 1000-4-6 specifies that harmonics and distortion must be at least 15 dB below the carrier level (-15 dBc). This can be checked during the test by using a spectrum analyzer with a directional coupler on the power amplifier output.

4.2.12 Power Gain

Amplifiers are usually specified to deliver their maximum power output for a given input level, typically 0 dBm. The power gain from input to output should be relatively constant over the whole operational frequency range. If it is not, then a higher level of drive signal is needed, typically at the edges of band coverage. This may place an extra demand on the output of the signal generator. The signal generator should have a maximum output level of +10 dBm to compensate for this.

4.2.13 Reliability and Maintainability

Despite the best efforts to improve reliability of equipment, power amplifiers do still fail. When this happens, repair and verification should be accomplished quickly. It is costly to have backup units available, and an amplifier that is out of commission will hold up a large part of the expensive test facility, not to mention product development.

4.2.14 Antennas

Antennas are frequently used in EMC work and their frequencies are summarized in Table 4.2. For radiated immunity, antennas may be uncelebrated if a field probe is used to measure the generated field. If antenna power is used to measure the generated field. If antenna power is used to calculate the generated field, or the antenna is used for radiated emissions measurements, then the TAF (transmitting antenna factor) and AF (receiving antenna factor) must be verified periodically by having the antenna calibrated. Linearly polarized antennas are preferred to reduce measurement uncertainty. Many times, two orthogonal antenna positions are used in successive test segments since the polarization for maximum sensitivity to radiated fields may be unknown.

Table 4.2 Representative EMC antenna Types

Frequency Range	*Receive Antenna*	*Transmit Antenna*
50 Hz – 200 Hz	Loop	Helmholtz Coil
30 Hz – 50 MHz	Road, Loop	Wire E-Field Generator
30 MHz – 320 MHz	Biconical	Biconical
100 MHz – 5 GHz	LPDA	LPDA
200 MHz – 40 GHz	Horn	Horn

The receive antenna factor is used to calculate the field strength of a signal impinging on an antenna. By measuring the antenna terminal voltage and applying the antenna factor, the field strength can be calculated. The receive antenna factor (AF) is defined as:

$AF = E - V - A$

Where AF = Antenna factor in dB/meter

E = Incident field strength in dBuV/meter

V = Voltage at input to EMI receiver in dBuV

A = Cable loss in dB

4.2.15 Field Measurement Probes

Field measurement probes must be calibrated if they are used to determine the field strength. Most probes will respond to the E Field component of an electromagnetic wave.

4.3 POWER MEASUREMENT

4.3.1 Thermocouple-type Sensors

These type of sensors are commonly used in the ships engine room and other machinery spaces. Thermocouples operate because dissimilar metals generate a voltage due to temperature differences at a hot and cold (ambient) junction of two metals. Since thermocouple sensors absorb the RF/microwave signal and heat the "hot" junction element, they give the correct average power for all types of signal formats from continuous wave (CW), to pulsed, to complex digital modulation, regardless of the harmonic content, wave shape or distortion of the signal. Thermocouple power sensors were the preferred sensor type for systems with complex modulation formats, as the sensor responded to the total aggregate power across its entire dynamic range. A simulated radar signal peak pulse power may be computed from the average power value and knowledge of the system duty cycle.

Thermocouple sensors typically have a dynamic range of only 50 dB, from –30 dBm (1 µW) to +20 dBm (100 mW). This restricted sensor dynamic range requires more time to measure the lower power levels.

4.3.2 Diode Sensors

These sensors are commonly used in tank gauging, flow gauging in cargo spaces and in a wide variety of uses in the engine room.

Diodes convert high frequency energy to DC by means of their rectification properties, which results from their non-linear current-voltage characteristic. A typical diode detection curve starts near a noise level of –70 dBm and extends up to +20 dBm. In yhr lower, "square-law" region the diode's detected output voltage is linearly proportional to the input power (V_{out} proportional to $V^2{}_{in}$).

Table 4.3 Power Sensor Types

Characteristic	*Thermocouple*	*Diode*
Sensitivity	Average	High
Dynamic Range	50 dB	90 dB
Setting Time	Average	Fast

4.3.3 RF Signal Generator

The RF signal generator may be the device that determines test frequency, unless an external counter is used. If the generator alone determines test frequency, this quantity needs to be calibrated. If the generator amplitude is not used for the field strength reference, the output amplitude may not need to be calibrated, so long as the field probes or power meters are calibrated and relied upon for amplitude determination.

Types of RF signal sources

- Swept -- sweeps over a range of frequencies, may be continuous or frequency "stepped"
- Signal generator—adds modulation, produces "real world" signal

Generators with internal modulation capability may be preferred over those that require an external modulator. Sine wave modulation is required for EC testing. Pulse modulation may be desirable to simulate radar signals.

- Range-range of frequencies covered by the source
- Resolution-smallest frequency increment
- Stability – frequency change with time or temperature
- Accuracy – how accurately can the source frequency be set
- Harmonic content. Some digital signal generators may have higher harmonic content than their analog counterparts.

Table 4.4 Signal Generator Characteristics

	Digital	*Analog*
Positive	Low Line Power Wide Frequency Range Good Stability	Higher Harmonic Content Inexpensive Less Phase Noise
Negative	Higher Harmonic Content More Expensive More Phase Noise	Requires More Line Power Less Table

4.3.4 Electronic Impedance Bridge

Another useful instrument in an EMC laboratory. This test equipment was originally designed to be used in the evaluation of antenna systems and has the following features:

- An RF signal generator that covers from LF to VHF, and UHF
- Frequency counter

- An inductance meter
- A capacitance meter
- Measures the SWR as a function of frequency

4.3.5 Spectrum Analyzer

An EMI receiver or spectrum analyzer is used for measuring emissions or viewing amplifier harmonics. The EMI receiver features a low noise floor and good frequency selectivity. In some cases noise floor and selectivity are traded for measurement speed and a spectrum analyzer is used instead of a receiver.

Table 4.5 Spectrum Analyzer Specification Sheet

Frequency Range	*10KHz to1 GHz*	*For typical models (Extended range models can measure much higher in frequency)*
Frequency Span	KHz to GHz range	User selectable
Resolution Bandwidth	KHz to MHz range	User selectable
Video Bandwidth	KHz to MHz range	User selectable
Signal Input Range	–100 dBm to 30 dBm	Must insure signal is below the maximum-to avoid potential damage
Demodulation	AM	Enables audio output of AM signal

The Spectrum analyzer screen displays amplitude versus frequency, but can also be made to display amplitude versus time. Amplitude measurements may be scaled linearly for showing signals near the noise floor, or logarithmically to show a large dynamic range of signals.

4.3.6 EMI Receiver/Spectrum Analyzer Detector Functions

Three detector functions are commonly used in measuring emissions. The first is the peak measurement. This measures the maximum value of the emission envelope at each frequency. The second type of detector function is the "quasi-peak" measurement. This yields a time-weighted peak value that is a function of the noise repetition rate. The assumption is that, up to a certain point, higher noise repetition rates are more annoying to the human ear. The last is the average detector, whose output corresponds to the average value of emission itself. The average detector is useful in measuring weak coherent signals near the system noise floor. (Both commercial and government requirements utilize these detector functions.)

4.3.7 Monitoring Equipment

EUT monitoring may employ digital, analog, or communication bus monitors, as well as audio or video links. These are usually connected to the operator's console via fiber optic cable so as not to perturb the field near the EUT. Fiber optic cable may penetrate the shielded enclosure via "wave guide beyond cutoff" pipes, whose diameter is determined by the highest frequency being utilized in the test chamber, and whose length is selected to provide at least as much attenuation as the shielding effectiveness of the shielded enclosure.

Monitoring devices may not themselves deviate under chamber test conditions. In the event of a EUT deviation, monitors that utilize a metallic connection to the wires of the EUT should be then disconnected to verify that they did not cause the deviation. The input impedance of the monitor should be such that it will not suppress an EUT deviation when connected.

4.4 ANALYSIS OF RESULTS

The test report may need to conform to certain standards, for example if the test is intended to demonstrate conformance to FCC requirements, the test report must contain specific information described as 47 CFR Part 2.1033. Standard ETSI EN 301 126 also specifies the format to which the report must conform. One reason for specifying format is that the information may be reviewed by different national regulatory agencies who may not share a common language.

If the report format is unspecified, the test report will at a minimum need to contain the information required by both the product regulation and any applicable quality standards such as ISO9000, etc.

Before beginning a test, an experienced EMC engineer should review the design, parts layout and packaging for best EMC practice.

The test group will need a meticulous, considered test plan. This will include the following items.

- Through description of EUT operation with schematic diagrams and layout diagrams
- EUT, wiring harness with connectors, and simulator, if required
- Monitoring equipment, fiber optic, video, etc., that will not affect the EUT immunity when attached
- The test frequencies (and levels for immunity) based on those most likely to cause the EUT to deviate behavior, or to interfere with radio services
- Operating modes in which to test the EUT based on analysis of which modes are most likely to fail immunity or emissions guidelines, or have the most severe consequences when they fail

- Production intent software, operating mode or modes most likely to exhibit a deviation or to maximize radiated emissions
- Information on whom to contact and how to contact them if questions arise

4.5 COAXIAL CABLES

4.5.1 Which System is More Efficient?

It is important to understand real transmission line characteristics and how those characteristics affect actual test setup or data from the testing. Let's look at two examples. The first example is shown in Fig. 4.7, where we have a 100-watt transmitter operating at 90 MHz, connected to an antenna by 70 metre of RG-58 coaxial cable. In Fig. 4.8 we have a 50-watt transmitter operating at 90 MHz, connected to the antenna by a 16 meters section of the same type of cable. Calculations show that the arrangement in Fig. 4.8 would perform better with more efficiency.

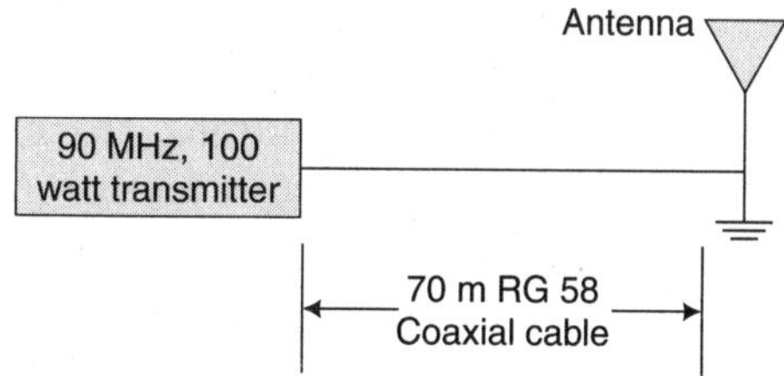

Fig. 4.7 High transmit power with high cable loss

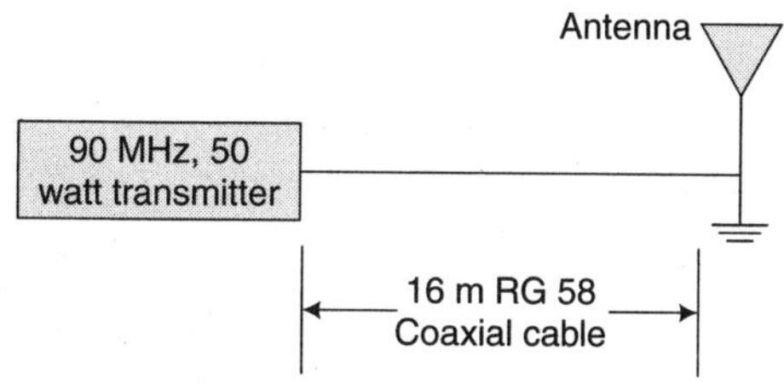

Fig. 4.8 Lower transmit power with less cable loss

4.6 EMC REQUIREMENTS FOR MARITIME ENVIRONMENT

Basically, there are 2 classes of EMC requirements that are imposed on electronic systems. They have to be electro magnetically compatible.

1. Those mandated by Governmental Agencies.
2. Those imposed by Product Manufacturers.

Government agencies are legal requirements. In the Maritime Sector, the Government has not yet imposed any EMC requirements for vessels in the maritime trade on a worldwide basis, excepting for certain Classification Societies, who do it on a selective basis.

Essentially, all the equipment on board vessels, come under the United States FCC requirements. They specify the range of frequency from 9 KHz to 3000 GHz. FCC requirement products are divided into 2 classes, i.e. Class A and Class B. Generally, the equipment used in the maritime field under the regulations is under Class A. (Annexure –1)

There are different standards developed for products, which are not under compliance of FCC. Some that are required in Germany are to be complied with VDE (Verband Deutscher Elektrotechniker). Those in France, require conformity to CISPR (International Special Committee on Radio Interference). There is no such specification required for vessels built in India, but most follow FCC requirements for installations and operations.

The requirements of those which do not come under the regulation of EMC control, are usually shown by random checking of (a) Radiated emissions, (b) Conducted emissions, including those for (a) Radiated Susceptibility, (b) Conducted Susceptibility, (c) Electrostatic Discharge.

In the Maritime Sector, there is equipment, which are used, that have severe design constraints for EMC. One is that, they suffer from the following:

1. Product cost
2. Product marketability
3. Manufacturability
4. Product development schedule
5. Diagnosis of the source of the problem
6. Diagnostic tools

Most Mariners do not think much of RFI. They usually view it as a technical problem. In most Ports, thousands of fishing vessels have radio equipment of various frequencies and they have a regulatory component that has social outcomes. Then there are amateur Radio Operators and the general public, which operates on FM at normal conditions. Usually, FCC standards of equipment take care of the problem. However, there are certain issues such as, (a) Power line Interference problem, (b) Leaky Cable Television problems, (c) Leaky Wireless Telephonic problems. (Annexure –2)

The interference caused by these are owing to (a) Noise (b) Overload (c) Unwanted Emissions. The European Union has got a regulatory scheme under Wheel Chart. ITU and IMO have also stipulated the requirements for compliance as under:

(i) Measuring noise levels at Listening Posts A.343(IX)

(ii) General requirements for Electromagnetic Compatibility (EMC) A.813(19)

Most of the equipment fitted on board, as far as communications are concerned, is usually certified.

During installation process circuits and components, filtering screen and shields, wiring, installation of cable trays, conductors in cables and grounding must all be done to strict specification. Usually such installations are done by convenience of space and economy of construction in vessel. Method of grounding cable screen and ground loops are usually specified.

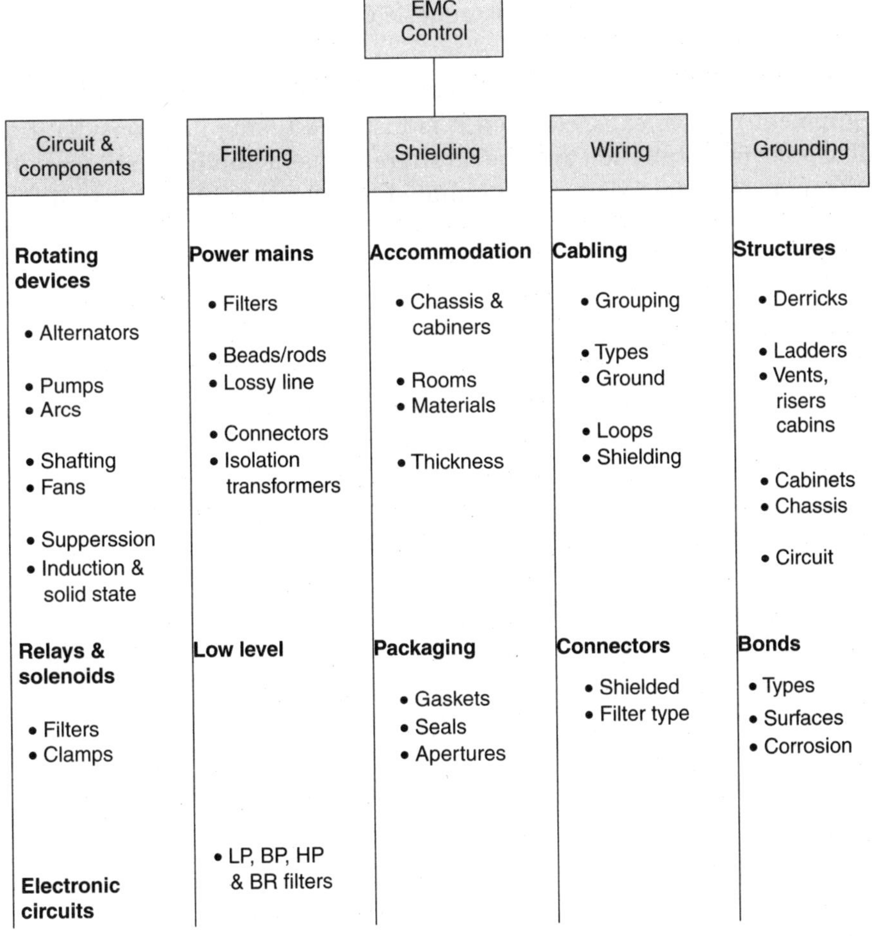

Fig. 4.9 An overview of available options to achieve EMC control

These are then adopted in vessels but adherence to standards and human errors negate all efforts by way of design. (Annexure–3)

(a) Equipment producing transient voltage, frequency and current variations is not to cause malfunction of other equipment on board, either by conduction, induction or radiation. (b) In distribution systems mainly consisting of (lighting/heating/ motors/ transformers etc.) and powered by synchronous generators, the total harmonic distortion in voltage waveform shall normally not exceed 5%. (c) In systems with large power consumption electronic power converters (or similar), the total harmonic distortion in voltage waveform shall normally not exceed 10%. (d) When above limits are exceeded, it is to be documented that no malfunction will be caused on equipment on board, in view of the actual existing waveform distortion. (e) All equipment is to be constructed to operate satisfactorily at the voltage and frequency variations, which can occur in normal use.

In general, there are Electromagnetic Compatibility Tests carried out. They are as follows:

The decision for testing of EMC is based on the interference to which the equipment may be exposed when it is installed taking into consideration all aspects of installation and the interference susceptibility tests specially the conducted interferences and radiated interference. (Annexure – 4)

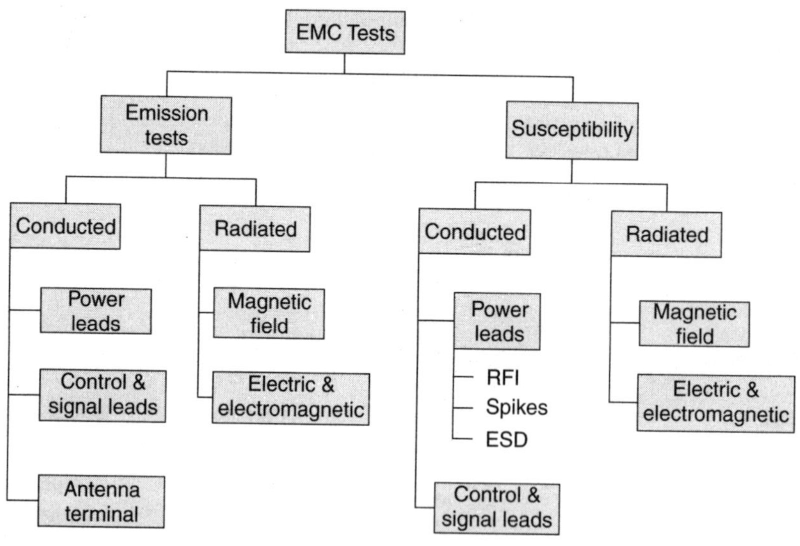

Fig. 4.10 An overview of EMC tests

Prior to embarking on a shipboard EMC test, an EMC check list is necessary. The status of the equipment is verified in preparation for EMC tests. The standards for tests are usually EU, BS or DNVC standards. The final step is the EMC management for which certain codes are adhered. This does not include distance related decoupling.

In the making of maritime equipment, especially, communication equipment, certain materials used are checked for radiation emissions and immunity. Conducted emissions are also checked. Usually these materials are checked in the manufacturing process of the equipment. This only covers a small part of the materials that are used for construction of the vessel. However, highly sophisticated vessels environments, communication gear vessels, vessels carrying extremely inflammable or dangerous cargo, are now-a-days been fitted with material, which have undergone its radiation tests. A typical process for performing tests is given in Fig. 4.11.

Performing Radiated tests

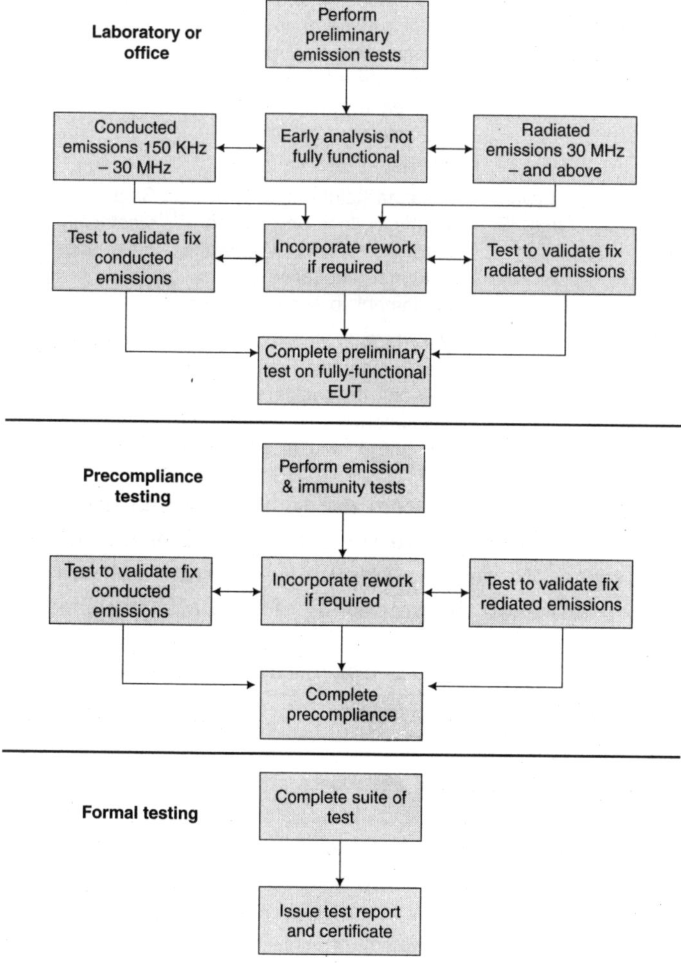

Fig. 4.11 Tests flowchart

As shown in Table 4.6 systems are classified by one of these characteristics into decoupling levels. The decoupling is given by a typical value. This value may be corrected according to the actual situation.

Table 4.6 Decoupling levels and typical characteristics (not including distance related decoupling)

Equipment location characteristics	*Decoupling (±10dB)*	*Typical EMC features*	*Examples*
No screening and no decoupling metal surfaces	Level 0/0 dB	Lines and cables have large antenna heights and loop widths; therefore little decoupling from each other and from the antennas	Ships mainly of wood plastic, or graphite composite.
Non substantial screening but decoupling metal surfaces	Level 1/15 dB	Lines and cables in contact with or closely above the metal surface have minimum antenna heights and loop widths, therefore decoupling from each other and from the antennas by at an average level	Ships with metal deck or superstructures with non-metal floor surface
Screening as well as decoupling metal surface	Level 2/30dB	Antenna heights and loop widths of lines and cables same as for level 1; average antenna decoupling of equipment to screened areas	Ship with metal deck and electrically conducting bulk-heads tanks
Closed screening or multiple screening e.g. by nested screens	Level 3/50 dB	Very high tightness against electromagnetic field; average antenna decoupling of equipment within closed screens or double screening	Submarine metal ships with several metal decks on top of each other

All the equipment of the system are to be classified into the following consequence classes according to the consequence of disturbance – This classification serves to set the immunity margins to characterize the valence of disturbances in the processing of the EMC analysis.

Table 4.7 Consequence of disturbance and immunity margins

Class	*Margin*	*Consequence of disturbance*
0	0 dB	No harmful effect
1	6 dB	Equipment for which disturbance of the function can lead to an additional load on the operators or limitation on the system efficiency
2	10 dB	Equipment for which disturbance of the function can lead to wounding damage to system or to restriction of efficiency of the system
3	20 dB	Equipment for which the disturbance of the function can lead to loss of life, to loss of the system, or unjustifiable restriction on system efficiency

Table 4.8 Some typical field strength values of material with reference to vessel frequency are given below.

Ship		*External equipment*		*Internal equipment*
Frequency	*dB uV/m*	*Reasons for requirement*	*dB uV/m*	*Reasons for requirement*
Metal 150 kHz to 1.5 Mhz	166	Internal MF transmitter (approx. 500 W) equipment approx.5 to 10 m away from antenna	130	Internal MF transmitter (approx. 500 W) 36 dB screen attenuation
Wood, Plastic 150 kHz to 1.5 Mhz	166	Internal MF transmitter (approx. 500 W) equipment approx.5 to 10 m away from antenna	160	6dB decoupling due to metal surfaces or installation outside high intensity radiation field.
Metal 1.50 MHz to 30 Mhz	160	Internal HF transmitter (approx. 1000 W) equipment approx 4m away from antenna	130	Internal HF transmitter (approx. 1000 W) 30 dB to screen attenuation
Wood, Plastic 1.50 MHz to 30 Mhz	160	Internal HF transmitter (approx. 1000 W) equipment approx.4m away from antenna	160	Internal HF transmitter (approx. 1000 W) no screen attenuation
Metal 30 MHz to 1 GHz	146	Internal UHF transmitter (approx. 100 W) equipment approx. 4m away from antenna	130	Internal UHF transmitter (approx. 100 W) 20 dB to screen attenuat-ion assumed
Wood, Plastic 30 MHz to 1 Ghz	146	Internal UHF transmitter (100 W equipment 4m away from antenna	146	Internal UHF transmitter (approx. 100 W) no screen attenuation assumed.
Metal 1 GHz to 40 Ghz	170	Internal radar transmitter, equipment outside of antenna main lobe and approx. 10 m away from antenna.	140	Internal radar transmitter, 25 to 30 dB screen attenuation assumed
Wood, Plastic 1 GHz to 40 Ghz	170	Internal radar transmitter, equipment outside of antenna main lobe and approx. 10 m away from antenna.	170	Internal radar transmitter equipment outside of antenna main lobe no screen attenuation

Within the scope of this analysis of all characteristics of the system power supply which are essential to the generation of propagation and effects of the disturbing pulse must be included and evaluated so that the couplings can be predicted. A typical example of an influence schematic and an example of estimated margin between noise and susceptibility is given in Fig. 4.12

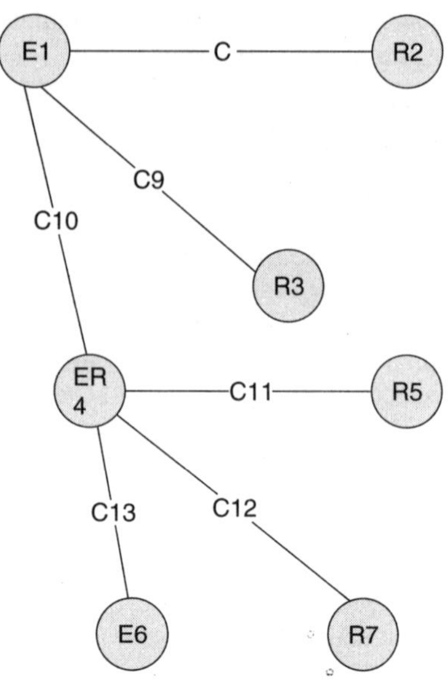

Fig. 4.12 Influence Schematic

Legend of Symbols

Rn EM receive i.e sensitive equipment; 'n' is the reference number

En EM emitter; 'n' is the reference number

Cn Coupling path; EM radiated signal; 'n' is the reference number

Cn Coupling path; EM conducted signal; 'n' is the reference number

Cn Coupling path; EM radiated and conducted signal; 'n' is the reference number

Example:

Emitter : MF radio transceiver: 500 W, 200 V/m

Coupling : Metal ship, distance between emitter and receiver 15m.

Receiver : Computer based system, EM susceptibility 10V/m

Parameter	*Emitter*	*Coupling*	*Receiver*
Decoupling		30dB	
Distance-dependent reduction		30dB	
EM platform	166dBu V/m	60dB	106 dBu V/m
Additional attenuation		0 dB	0 dB
Consequence class			10 dB
Required Susceptibility			116 dBu V/m
Specified Susceptibility			140 dBu V/m
Margin between noise and susceptibility			24 dBu V/m

The final document would be the EMC management sheet as appended, which is compiled. This indicates, if all the EMI has been taken care off or is at dangerous levels of radiation and EMI are still prevalent. The error owing to EMI of internal environment effects in HOE exists strongly. **(Annexure – 5 & 6)**

4.7 TROUBLE SHOOTING CONCERNS

One of the important issues faced by Mariners out at sea are equipment which once compliant becomes non-compliant in the field or causes problems with other electrical equipments, owing to emissions or immunity. The complexity of the coupling, which can give rise to such problems, is illustrated in the Fig. 4.13 below:

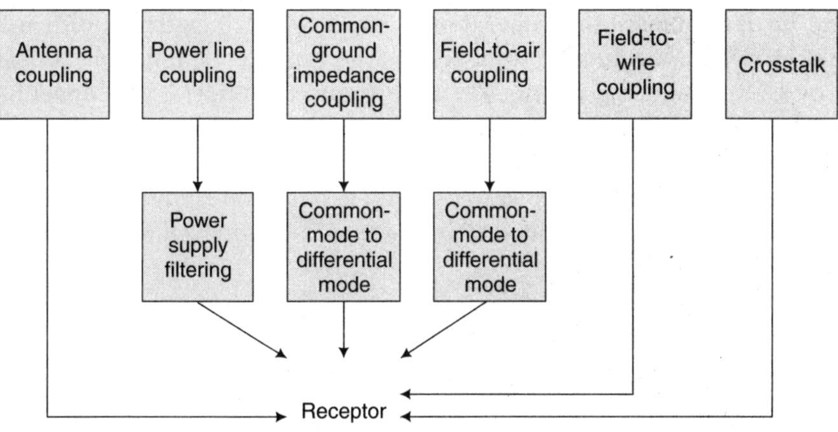

Fig 4.13 Complexity of coupling

The measurements of such couplings as well as generated noise on 10 year old ships are given in Annexure-7. It shows how repairs, wear and tear and use of non-standard material including faulty installation has caused the unusually high radiation levels.

Chapter 5

EMC Modeling

5.1 EMC MODELING

Performing EMC Tests is a demanding process and they are conducted late in developmental or operational state, when it becomes difficult to correct EMC Modeling. One of the method of overcoming this problem, is by EMC Modeling. In the Maritime Industry, lately, EMC Modeling is being experimented upon new vessels designed and under construction. A fair test has been made of the EMC problems on existing vessel and both internal and external sources. This is based on Emissions Modeling. A possible scenario for modeling in the maritime environment is given as under:

5.1.1 The Value of EMC Modeling

Performing EMC tests is a time-consuming, resource-demanding process. Unfortunately, these tests are often conducted late in a developmental stage, when correcting an EMC problem can be even more difficult and expensive. Validated analytical and numerical methods have the potential to become increasingly important as a process to determine effects of external fields on a car's electronic systems, or anticipating how emissions will develop.

Proper electromagnetic compatibility modeling can reduce development time and the accuracy of the modeling results can be sufficient for planning the layout of cables and components. If the analysis is made during early stages of vessel design, when layout is more flexible, changes to improve immunity can be adopted without appreciably increasing costs. Early simulation and testing can significantly reduce the time spent testing the final product in the chamber.

Resonances inside a vessels enclosed spaces enhance incident fields, which can even rise to several times the external value in some vessel locations. Therefore, the electronic system components (wiring harnesses end electric and electronic devices) must be designed to be immune to such disturbances. Knowledge of these locations can be used to locate cables and electronic systems in regions characterized by reduced field intensities. Conversely, knowledge of vessel locations that couple well to outside environment can be used to advantage for locating devices like RKE modules with integral antennas that are designed to couple to devices outside the vessel.

Modeling can be used to study the coupling of both internal and external sources. The effects of in vessel portable radio transmitters and both intentional and unintentional receivers (i.e., cellular phones, CD players, GPS, "handi-talkies", etc.) can be as significant as sources outside the vessel.

The challenge is that EMC modeling is a difficult task due to the nature of EMC itself. Since EMC is the study of "things not on the schematic", much of the complexity in EMC modeling involves determining the relevant parameters to model. Unlike circuit analysis, which is more defined, modeling for EMC is a discipline that is still being designed and developed.

Modeling that has been done to date has primarily evolved around printed circuit boards and component level devices. These devices are more concise and easier to define than the large systems (such as completing a model for an entire seagoing vessel or system). A search of the references and literature available shows that the techniques have been used with some success on many problems such as cross talk on a computer board or within a digital device.

The goal of vessel system EMC modeling is to enable efficient and effective analysis to augment or to completely replace time consuming and expensive testing. This is where modeling has the highest likelihood of impacting shipboard system EMC work and where the most significant benefits are expected.

5.2 EMISSIONS MODELLING

In order to understand the basics of EMC modeling, it is instructive to look at the various areas that have been covered and the specific items in those disciplines. One of the first areas of modeling was to anticipate the levels and types of radiated emissions that could be expected from a device or component.

The reason that this was undertaken first was:

1. There have been numerous studies and data published for years about the occurrence of RE (Radiated Emissions), since it was one of the earliest aspects of EMC.

2. The phenomena responsible for RE are relatively understood. These have been discussed in other portions of this text, and are essentially due to loop areas, common and differential mode current, etc.
3. Many of the actual device and pcb (printed circuit board) layout characteristics that contribute to RE are well documented in the literature.

The task of modeling for immunity is more difficult for a number of key reasons.

These include:

1. The requirement to understand component operation and how it may be affected by external sources of energy (which varies from device to device)
2. The coupling path and efficiency of that path in the development of the immunity problem.
3. The manner in which other PCB components modify the interference before it reaches the susceptible component.

5.2.1 Modeling Challenges

Given the need to understand and correctly define the physics of the circuits and the systems, there is still a need to be able to develop the exact numerical tools to define the energy coupling paths by:

1. Understanding the path(s)
2. Qualifying the paths
3. Identifying the applicable corrective actions or countermeasures that can minimize undesired interactions

From our discussion on printed circuit board layout issues, it should be recognized that the board design and layout are important for the following reasons:

1. The need to understand the characteristics of PCB trace and board interaction and coupling metrics.
2. The need to define and quantify items such as power bus noise that may exist which can contribute to R1 problems due to the impact of adding additional energy to a circuit, which already has EMC issues.
3. The requirement to clearly know and define the parameters of the passive elements (lumped and distributed).
4. The need to define various system or component structures that may act as a shield to the external energy. (As we have seen, shielding can substantially change EMC performance characteristics.)

Other contributors to the modeling challenge are the impact of the types of wires and cables that are used, how they are bundled together, and their construction. System or component structures such as frames or cases may actually be acting as energy conductors. This also needs to be considered in the model-this may be a complicated task.

5.3 GOAL OF MODELING

The overall goal of EMC modeling is similar to the modeling work that is done in any other aspect of engineering. The desired approach is that the modeling will assist with key steps in the validation process. One item is the development of the circuit layout. The process of board or circuit layout many times does not include issues that are important to EMC performance. The major contribution of modeling would be to identify potential EMC issues. This is also known as the 80 /20 rule. The application of this rule to EMC is that 20 percent of the items cause 80 percent of the problems. Modeling should contribute to reducing the work that is expended in trying to identify 100 percent of the issues, of which 80 percent will be determined not to be problems. It is the goal of modeling that the 20 percent of the issues will be able to be clearly defined in terms of their significance to EMC

Another desired aspect of modeling is the ability to define a model based on a small component, then being able to "upscale" this component to a larger complex system level analysis. This is important to comprehend as it can mean the difference an accurate system model and results that may be inaccurate.

The intention of EMC modeling should be to take the component level data and use this data to increase the accuracy of the system analysis.

Some "thought starters" with this point include the following.

1. Use the model to create a "transfer function" from the component level information, then use this transfer function at the system level.
2. Use the model to develop knowledge about system impedance.
3. Be able to model the current flow paths and to quantify the magnitude of the common and/or differential mode currents.

A possible Process represented in the diagram information flow paths shown in Fig. 5.1.

5.3.1 Vision for Modeling

What should be the long approach for EMC modeling? There are some issues that appear to be common between the work and processes use in the

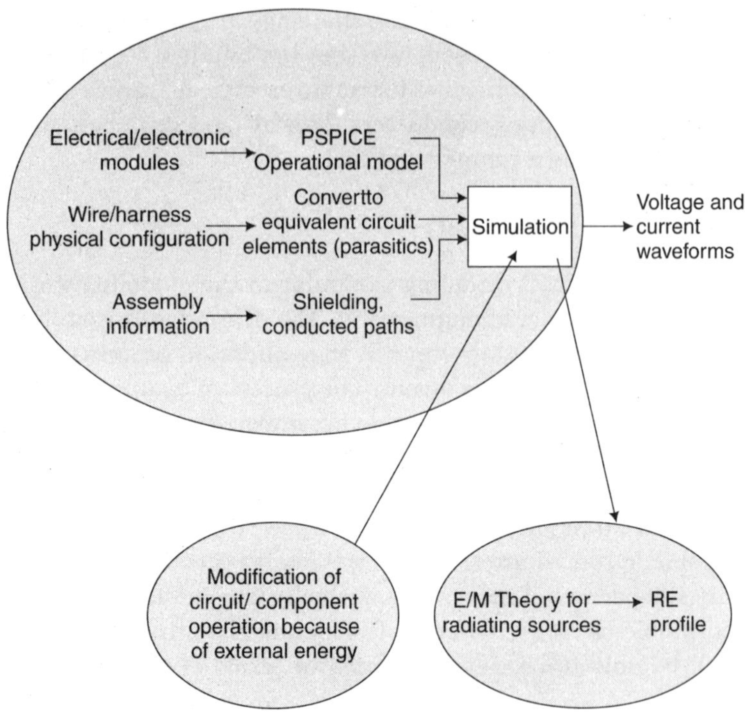

Fig 5.1 Model inputs and outputs

studies that take place in the computer engineering work. These include the study of "signal integrity" to understand how power and signals are transferred through a circuit.

The process of EMC modeling resembles the following:

1. Use modeling for initial look at system performance.
2. Conduct focused testing on those areas identified as critical, or do not have good boundary conditions associated with them.
3. Provide feedback on the integrity of the modeling results to continue to make further improvements to the model

5.3.2 MODELING/PREDICTION TECHNIQUES

While prediction methods are available for predicting RE, there are complicated programs which require tedious circuitry input parameters. More general calculation routines are available in which the calculations are done with pencil and paper or computer spread sheet. Table 5.1 gives an example of the general type calculation and presents a series of columns which are added to determine the predicated RE from a given component. Basically, the Fourier transform is computed for a given signal. The amplitude (dB µV) is placed in column 1. A correction factor is added to account for the conversion of the conducted data to free space. This factor is assumed to be –34 dB for a perfect λ/4 the frequency, f3, is computed to give a factor that takes into account the actual length of the cables. The calculation is shown in Table 5.1 and corrects the λ/4 assumption. Finally the number of leads from the signal is accounted for. The measurement distance is also factored in; for most test applications it is 1 m and requires no correction factors. The resulting factor is compared to the specification at that frequency to determine the dB of attenuation needed.

The attenuation is obtained by using metal equipment housing, shielding, twisting, etc. This prediction method is reasonable for checking the most likely culprits in the circuits (diodes and transistors that switch high levels of current).

Table 5.1 RE prediction Analysis

Frequency	*(1) Cn*	*(2) Voltage to E Field*	*(3) Antenna Factor*	*(4) Number of Leads*	*(5) Distance*	*(6) Result*
20 MHz	130 dB μV	–34	–2.7	6	0	99.3 dB μV

1. Frequency domain amplitudes – Fourier Transform
2. Voltage to field intensity level 1 meter away from conductor = –34 dB
3. –10 log (f_3/fx) where f_3 ** 3 $(10)^8$/4L, L = wire length in meters when fx/f_3 the correction factor = 0 dB.
4. + 10 log N, where N = number of leads
5. –20 log D, where D = test distance in meters
6. Result in dBμV/m, sum (1) through (5).

The above Table 5.1 is shown as a flow chart in Fig. 5.2.

Modeling can be used to help formulate an EMC test plan, by predicting which frequencies couple to wiring harnesses and modules. It can be used to decide whether additional isolation is needed or not, thereby letting the design engineer know whether to budget for shielding components or twisting wiring harnesses. It can even be used to help orient modules, for example, Remote Key Locks or Remote Start with built in antennas for maximum coupling to external sources of energy, when desired.

5.4 SAFETY INDEXING OF SHIPS

Safety indexing of ships is a comprehensive way of calculating the safety of vessel. This is more risk based than rule based and its basis is on the understanding that

Total Risk = Risk owing to Reliability of machinery X Risk owing to Human and Organizational error

Consequently $R1 = R1_{RISK}\, X\, R1_{HOE}$ or

Safety index $SI = 1 - RI$ and exponentially put as $= 1 - e^{-3R+1}$

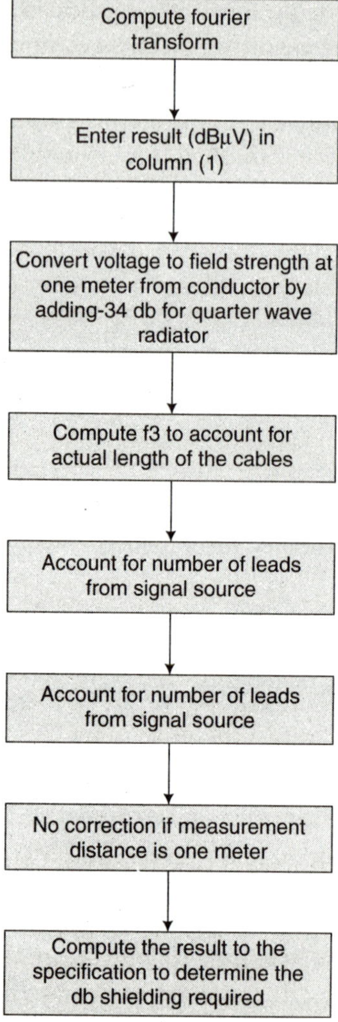

Fig. 5.2 RF prediction analysis

One of the most important constructs of safety indexing is understanding and mitigating the EMC environment and ensuring pollution to that effect is least. There are now many methods of EMC management evolving and they would differ in approach for vessels which are existing, or for new construction as well as for its nature of operations.

Electromagnetic interference is a form of environmental pollution. Specialized circuits are becoming increasingly vulnerable to pulsed electromagnetic interference, such as ESD, EFT, SURGE, and power quality failure transients. These are both visible and invisible, and range in intensity from lightning bolts to normal switch arcing, to small but potentially disastrous electrostatic discharge from the human touch or even furniture. The sparks involve very high power or extremely fast transients, thus effecting quality of ac power mains, DNV Standards.

In fishing vessels, this frequency spectrum pollution emanate out of

1. Electrostatic discharges in nature
2. High voltage cabling
3. Radar's
4. Electric motors, actuators
5. Ignition system of engine
6. Wireless and communication systems
7. Marine communication antennas
8. House keeping appliances, cooking appliances
9. Lamps of special types
10. L.V. fuses and H.V. fuses
11. Static watt-hour meters

The effect of EMI can range from minor nuisances which are easy to list and tabulate to some serious consequences like communication failure, power failure, shocks, navigational instrument failure. The number of cases in which unexplained power failures, communication equipment failures, shock and other major effects to safety of vessels and personnel were analyzed from casualty investigations.

The investigations attributed all such effects to environmental factors. Results of the investigation revealed that

1. All 112 cases of casualties, had some input of EMI.
2. Questionnaires circulated to 100 individuals including vessel owners, to ascertain the unexplained nature of such effects and the results of same are given below
   (a) Shocks to personnel and radiation effects from equipment of major nature-08% of cases this resulted in casualties.
   (b) Communication failures, wireless equipment failure-12% of cases without any breakdown of equipment.
   (c) Navigation equipment failure-05% of cases without any breakdown of equipment.
   (d) Other major effects-07% of cases.

Total of 32% of cases.

The experts committee concluded that such unexplained causes were EMI related. None of the vessels meet any EMC standards. The navigational and communication equipment are considered adequately protected if they meet EN 50000-1-2-3 or DNV standards.

The experts lay down that the highest level of safety would be a total compliance with same 50000-1-2-3 or DNV standards. Anything less than total compliance would be a percentage in terms of equipment compliance.

The terms of reference were in compliance of

1. Shielding
2. Electromechanical devices
3. Ferrit components
4. Cabling
5. Grounding

The equipment were classed in the following manner,

Table 5.2 Classification of equipment as per radiation levels

Class	*Margin*	*Consequence of disturbance*
0	0dB	No harmful effect
1	6dB	Equipment that have some consequences and harmful effect
2	10dB	Equipment which can lead to injury, effect on safety
3	20dB	Equipment with major effect on personnel, vessels safety

Thus there are 22 equipment on a type of fishing vessels with following class specifications.

Table 5.3 EM compatibility analysis

Class	*Ideal Vessel*	*Vessel A*	*Analysis*
0	12	10	Nil
1	6	3	50%
2	2	0	0%
3	3	0	0%
3	11	3	27.8%

Most vessels were below 30% compatibility. At 32% of casualty attributed to EMI and EMR the balance of environmental effects were 68%.

The gradation used for evaluating percentage of compliance was

100%	– 0
70% to 100%	– 1
30% to 70%	– 2
0 to 30%	– 3

The same was used as inputs into the safety index structuring. The basis for accepting 100% compatibility was by either

1. Manufacturers self declaration
2. Third party certification.

Having assessed accident data, casualty reports, and taken into account experts view in judgmental analysis as inputs into index, the critical error factors in risk, reliability index formulation and human and organizational index formulation is given in the next chapters.

Chapter 6

Shipboard Electro Magnetic Environment

6.1 SHIPBOARD COMMUNICATION EQUIPMENT

A shipboard radio station includes all the transmitting and receiving equipment installed aboard a ship for communications afloat. Depending on the size, purpose, or destination of a ship, its radio station must meet certain requirements established by law or treaty. For example, large passenger or cargo ships that travel on the open sea are required by the Communications Act and by international agreements to be equipped with a radio station for long distance radio communications. Passenger ships that travel along the coast must be able to communicate at shorter range with coast stations. These are examples of "compulsory ships" because they are required by treaty or statute to be equipped with specified telecommunications equipment.

Smaller ships used for recreation (e.g., sailing, diving, fishing, water skiing) are not required to have radio stations installed but they may be so equipped by choice. These ships are known as "voluntary ships" because they are not required by treaty or statute to carry a radio.

The marine radio equipment listed below may be used aboard a ship. If the ship must be licensed, all equipment is authorized under a single ship radio station license.

VHF Radiotelephone (156-162 MHz) - Used for voice communications with other ships and coast stations over short distances.

Radar - Used for navigating, direction-finding, locating positions, and ship traffic control.

EPIRB - Emergency Position Indicating Radio Beacons, or EPIRBs, are used when a ship is in distress, to emit a radio signal marking the ship's location. Extreme care must be taken to prevent inadvertent activation and batteries should be replaced prior to expiration date.

Single sideband Radiotelephone (2-27.5 MHz) - Used to communicate over medium and long distances (hundreds, sometime thousands of nautical miles).

Satellite Radio - Used to communicate by means of voice, data or direct printing via satellites.

Radiotelegraph - Used to communicate by means of Morse code facsimile or narrow-band direct-printing.

Survival Craft Radio - Used for survival purposes only from lifeboats and rafts.

On Board Radio - These are low-powered radios used for internal voice communications on board a ship or for authorized short range communications directly associated with ship operations. In addition, ships may use GPS or LORAN receivers, depth finders, citizens band (CB) radios, or amateur radios (an amateur license is required).

6.2 SHIPBOARD TEST AND EVALUATION PROCEDURES

At –sea in a deployed environment, much more care and planning must be given to the required cabling and electronic power requirements. Approved techniques for securing cables and a dedicated plan for data acquisition and processing are required. Furthermore, at sea, testing is much less restrictive of emitter power levels and less prone to industrial EMI. Conversely, in-port testing is easier to schedule, is more flexible to spontaneous test modifications, is less affected by adverse weather conditions, and can use an onshore wall-powered vessel as the remote site for data acquisition and processing. However, in-port transmitted power restrictions and industrial EMI can limit the scope of the testing. The ideal scenario is to perform preliminary in-port tests which culminate in an at-sea demonstration.

Once configured, the actual shipboard tests which are to be performed will be similar for both in-port and at-sea testing. The general test procedure will be to place the remote photonic probe at each of the selected topside positions and individually and simultaneously excite the probe with the shipboard transmitters. The shipboard emitter's frequency and transmitted power level information will be carefully measured. The E-field strengths, intermodulation product levels, and modulation characteristics will be recorded at the monitoring sites, with the transmitters operating in their normal operation mode (i.e. power levels and modulation format). Performance degradation due to industrial and military EMI caused by external off ship emissions will be noted. Data acquisition and recording will occur in an approved topside site (or inside an onshore vessel) through a length of 300 meters of single mode optical fiber. Hand held two way radios will be used to coordinate between the transmitter rooms and the processing site. Received power levels will be recorded from which electric field strengths will be deduced. The electrically powered laser and optical detector, as well as the required processing electronics, will be assembled at the remote processing site, where standard single phase 115 V AC power which can supply a current of 20A is required.

Upon completion of the shipboard testing, the data will be analyzed and an evaluation will made of the photonic EM field probe for shipboard EME monitoring. A report will be submitted which will include the shipboard test results, the final photonic EM field probe configuration, recommendations for further developments, and lessons learned for future developments.

6.3 SHIPBOARD EME DETERMINATION

When available, experimental weather deck EME data are used for estimating field strengths at different topside locations due to the various shipboard emitters. When experimental EME data are not available, a far-field approximation is used. In the far-field, the main lobe weather deck EME due to a given emitter can be calculated using

$$Pr = (P_t)\,(G)/4\pi R^{2;}$$

Where

Pr = Received power density (W/m2)

Pt = Emitter level (W)

G = Antenna gain (dBi)

R = Radius (m), antenna to sensor

and the electric field strength (V/m) is then given by

$$E = (\eta P_r)^{1/2}$$

where $\eta = 377\ \Omega$ is the free-space radiation resistance. When experimental data are not available, this expression is used (with gain correction for directional antennas) for estimating the field strength due to a shipboard emitter at various topside locations. For the higher frequency transmitters, this approximation should give fairly accurate results.

6.4 RESULTS

(a) Broadband electromagnetic (EM) field detection using antenna-coupled fiber optic links has been successful demonstrated in the2MHz to 18-GHz frequency range. This remote sensing system is potentially operable from 2 MHz to 50GHz and can be packaged into small, lightweight units. Both amplitude and frequency information of multiple radio frequency and microwave signals have been simultaneously monitored with this wideband optical system. A root mean square electric field sensitivity of approximately 1 V/m and a spurious free dynamic range of 109 dB/Hz2/3 have been demonstrated with the18-GHz externally modulated system.

(b) Semiconductor optical waveguide modulators have been developed as an alternative to lithium niobate Mach-Zehnder modulators for this application. Systems employing the less mature semiconductor optical modulators have shown greater modulation efficiency for a given bandwidth at the expense of added optical insertion loss than lithium niobate based systems.

(c) Remote modulator bias control techniques have been investigated and systembased on the active control of the modulator direct current bias to compensate for bias position drift has been developed. This system utilizes a remote power-by –light approach in which a low frequency optical control signal is sent to the modulator via as optical control signal is to the modulator via an optical fiber and is remotely monitored to properly bias the optical modulator. This system has been successfully demonstrated in the 10°C to 90°C temperature range.

(d) A remote optical polarization controller has been successfully demonstrated, which eliminateS the need for expensive polarization maintaining fiber. The controller has an optical insertion loss of 4.5-dB, resulting in a 9-dB penalty in detection sensitivity. Recent improvements in liquid crystal technology have resulted in polarization controller units with <2-dB optical insertion loss.

(e) Small spiral antennas with responses from 300 MHz to 50GHz have been demonstrated and are ideal for this shipboard emission monitoring system. Below 500 MHz, a number of approaches are currently being investigated to optimize the size/gain tradeoff characteristics of various antenna structures. One approach is to fabricate slightly larger spiral antennas to enhance the low frequency response. A second approach to increase the low frequency response is to fabricate spiral antennas on a high dielectric constant substrate. A third approach has been to use a sectoralized loaded monopole antenna. A fourth approach is to suffer the gain penalty and use an electrically short dipole antenna. Each technique has its advantages and disadvantages, although the use of an electrically short dipole antenna is the simplest approach, and results indicate that acceptable 2-MHz to 500MHz electric field sensitivities can be obtained using this approach.

6.5 SHIPBOARD DEMONSTRATION TEST PLAN

The purpose of shipboard demonstration is to qualify the technology under development and provide ship operators with an example of the enhanced capabilities it can bring. The purpose is to temporarily mount the prototype sensor assembly at various topside positions while controlled shipboard emissions are recorded. Details that need to be addressed include ship class, measurement sites, test and evaluation procedures, and documentation plans. This is a accepted procedure before the actual recording is validated.

6.6 SHIP CLASS SELECTION

A number of surface ships have been considered for use in the shipboard testing of the EME monitoring system. Among these, the ocean graphic research vessels and fishing vessels class are considered the top candidates. Important factors influencing ship selection are the number of ships in the class, the EM configuration of the ship, and the availability of the ship class for shipboard measurements. The existing ship class experimental EME database and the past experience of associated personnel in making shipboard measurements also weighted heavily in ship selection. Cargo ships, ships with very high levels of automation, ships having unapproved equipment are specially selected for such tests.

Chapter 7

Effects of Cabling on Ship

7.1 CONDUCTED EMISSIONS AND IMMUNITY

We have discussed one of the items of an EMC model that consists of the coupling path being radiated through the air or the vacuum of space. This chapter will discuss the fact that is important to remember that emissions and immunity problems can also result from being conducted along some type of wire, cabling, or even conductive portions of a vessel or system assembly.

7.2 INDUSTRY EMC APPROACHES

Why are cabling and harnessing important to the shipping industry? The reason is that the industry recognizes that EMC issues can occur on vessels with electronic modules interconnected by wiring harnesses.

Cabling as interconnections is still very common in the shipping environment. Shipping systems for years to have relied on wiring and harnessing to provide power and signal distribution throughout the vessel. It is not anticipated that this will change in the near future; therefore, coverage of effects of all cabling and harnessing is discussed in this chapter. Key elements to understanding harnessing and cabling are the parasitic inductance and capacitance.

7.3 SIGNIFICANCE OF WIRING TO EMC

Let's look at the frequently overlooked, but key items in EMC. In EMC most of the attention is placed upon the obvious components and systems and it is assumed that these are responsible for all the EMC issues. Frequently the connections between the systems and components are not considered, or looked up on as "merely wires", with no particular reason to investigate any deeper into their characteristics. Unfortunately cables and wiring systems can play a large role in EMC issues, and can be the source of many EMC problems! This is because although the wires and cables are "passive devices" on their own, they can create "parasitic" capacitors or inductors that are a result of wires or cabling can cause problems just as if they were intended components included into the circuits. This chapter will discuss the EMC aspects of wiring and cabling, and this relationship to helping minimize EMC issues or unanticipated development of EMC issues.

7.4 ROLE OF WIRING IN EMC

The key elements to understanding how wiring and cabling may contribute to EMC concerns follow. When we connect a component or system to wires, our intention is to bring energy in the form of power or signals into and out of the device or component itself. This connection can also result in the conduction of unanticipated energy or noise into or out of the device. There are two primary types of current that we need to understand for their contribution to EMC:

- Differential mode current
- Common mode current

Differential mode current is a current that we intend to have, such as the power supply and the power return current flowing on two different wires. This current will flow in opposite directions on each wire. Mother example of differential mode current is the signal current and the signal return current flowing along wires line in opposite directions. Sometimes we may have the situation where current flows in several wires or conductor that flow in the same direction along multiple paths. This is called common mode current. The challenge is that both differential and common mode current can cause EMC problems. Each of them has its unique EMC characteristics, including conducting noise along the wire, which acts as an "antenna" to receive external energy or to cause energy to be radiated from the system.

7.5 EARLY VESSELS WIRING

The impact of shipping wiring harnesses contributing to EMC issues on today's Vessels can be significant. If we compare earlier models of vessels, there were only a few electrical and no electric devices (except for the radio-if there was one installed). If we look at the vessel wiring, we notice that the wires were primarily connected to only switches to control various lights and the ignition system. These vessels had approximately 50 meters of cable, and the wiring weighed only about 4 kgs. If we now contrast with a recent model vessel, we can see that there is a significant difference in both the number of items on the Vessels and the complexity of the wiring companies they harness. Today's vessels can have well over a few kms of the cable that may weigh over a few tons. The complexity from the EMC standpoint is that, in comparison with the older vessels, today's vessels have several characteristics. This is because:

- We did not have many electrical or electronic devices
- The devices that we did have were connected by relatively simple harnesses
- As a result, there were minimal EMC issues due to wiring

7.6 COMMON MODE AND DIFFERENTIAL MODE CURRENT

A key concept in EMC is understanding when current is flowing and where it is not, if this sounds like a basic statement.

There are two types of current that can flow in a wiring harness. At this point it is important for us to examine some of the characteristics of these two types. There are called "common" and "differential" mode currents. Common mode current means:

"Current flows along two or more conductors in the same direction at the same time". This is shown in Fig. 7.1.

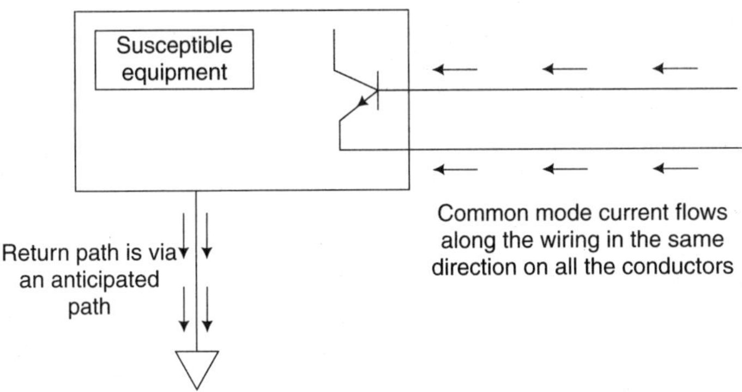

Fig. 7.1 Common mode current

7.6.1 RF Emissions and Immunity

In "differential mode" current flow, it means that the current flow is 180 degrees out of phase with other current(s).

This is identified as I_d, and the common mode current is identified as I_c. In a pure differential mode current flow, I_d would be equal and 180 out of phase with each other. In practice, the total common mode current would be the sum of both of the I_c shown on the diagram. This would then result in a total flow on each of the lines of I_1 and I_2. these would be the algebraic sum of I_d and I_c.

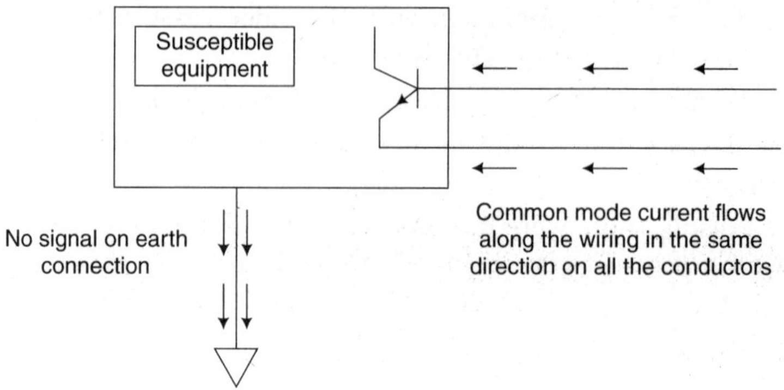

Fig. 7.2 Differential mode current

While seeming to be insignificant in themselves, a study of common and differential mode current can assist in diagnosis of EMC problems. This is because each of the currents creates different types of conditions. Ideally, our circuits and systems would have differential mode current, with clearly defined paths. This is not always the case, and we have experience with common mode flow in many instances.

The advantage of differential mode current is that it follows predictable (and the intentional) current path. Common mode current will follow the path of least impedance, even if that path results in unanticipated system or component operation. This can be even a result of current flowing into a "output" that was planned. If this does occur, there may be paths of current that cause incorrect operation.

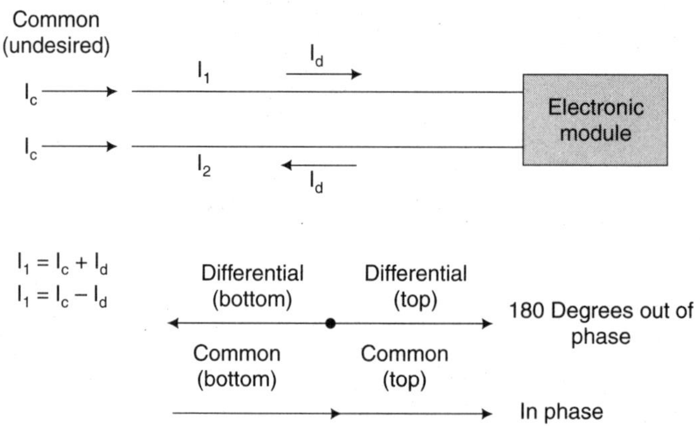

Fig. 7.3 Differential and common mode current

The following is representation of the difference between common and differential mode current.

Another aspect of differential mode current is that if the magnitudes are the same and 180 degrees out of phase, the magnetic fields due to each current flow will cancel. The result is that if we integrate in the region about the total conductor, the magnitude of the magnetic field will be equal to zero.

Let's look at the radiated emissions that would occur from each type of current flow, and how to minimize their impact.

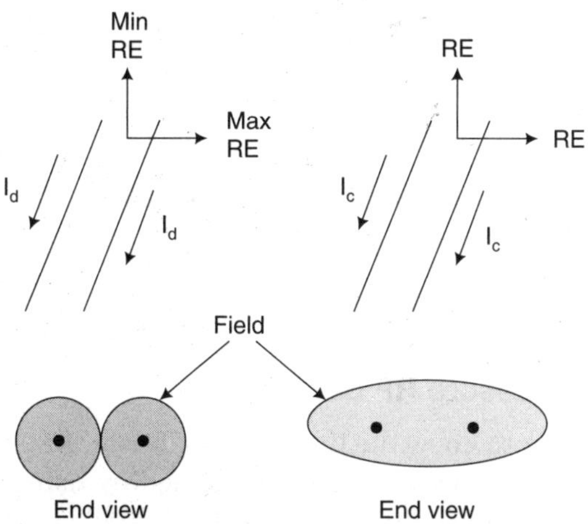

Fig. 7.4 Differential and common mode radiation

Table 7.1 Common Mode and Differential Mode RE

	Correction	*Field Intensity*
Differential mode Sensitive	Reduce Current	Not Rotation
	Reduce Line Length	
Common mode Sensitive	Reduce Current	Cable Rotation
	Reduce Loop Area	

Table 7.1 shows some of the radiated emissions characteristics of both common and differential mode current. If we look at a diagram of these current types, we see the following:

For the DM current, the minimum RE field occurs at 90 degrees from the plane of the conductors, and the maximum RE field occurs in the plane of the conductors. An end view shows a sideways figure "8" that surrounds the conductors.

For a loop of less ¼ wavelength for the DM current, the minimum RE field occurs at 90 degrees from the plane of the conductors, and the max RE field occurs in the plane of the conductors. If the loop length is ¼ wavelength, the pattern changes by 90 degrees

In the case of the CM current, we have a symmetric RE field that surrounds the conductors. This is due to Ic, and acts similar to the way an antenna operates (CM current is some times called antenna current due to its efficiency of RE). The end view for CM current looks like a loop around the conductors.

Let's summarize the characteristics of RE Caused by RF Current

Table 7.2 RE caused by both DM and CM currents

Type	*Effect Analysed by*	*Caused by*
Differential mode	Transmission Line Models	"Desired" Current
Common mode	Various Effects	"Undesired" Current

Table 7.2 illustrates that we can have RE caused by both DM and CM currents, and the analysis of the fields from each of those is different. This also illustrates that the CM currents that cause RE can be difficult to analyze, another reason we don't want those types.

7.6.2 Ways to Measure RF Current

Since it is important to know the flow of noise or RF current within a circuit or along a conductor, this can be accomplished by using various types of probes. These probes typically are clamped or installed around a conductor, and measure the RF current to the magnetic field that is created around the conductor (right hand rule). The probes can be calibrated to provide either relative or absolute indications of the current. They are installed as shown in Fig. 7.5 and would give the results as shown:

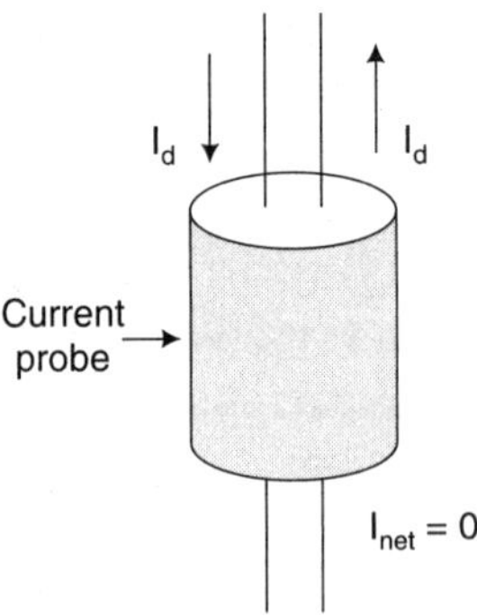

Fig. 7.5 Measuring differential mode current with a probe

From figure 7.5, it can be seen that that the DM current is surrounded by the probe. Each of the conductors of the DM current will produce a magnetic field, and each magnetic field is the opposite direction from the other. This will result in a current probe measurement of zero, since there is DM current.

In the case of measurement of CM current, shown in Fig. 7.6, the measured value will not be zero, since it is based upon the algebraic sum of each of the magnetic fields, which is based up on the current flow.

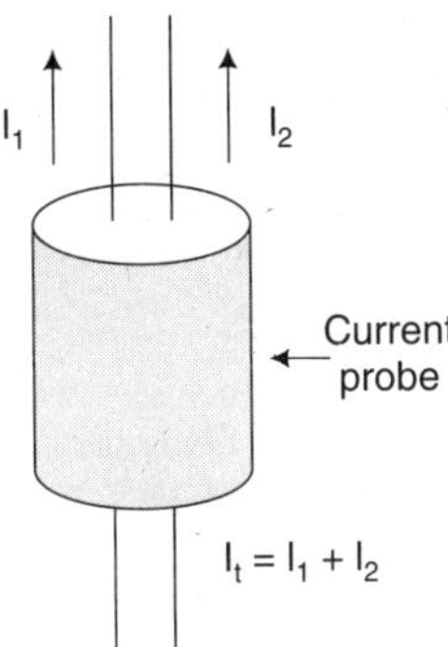

Fig. 7.6 Measuring common mode current with a probe

This illustrates that use of current probes can be an important diagnostic tool, provided that it is understood how they operate and what type of measurement is obtained.

Figure 7.7 shows a practical example of the use of a current probe for RF frequencies is in verifying the functionality of a "group strap". These types of connections are used frequently in an attempt to reduce emissions from a component or system, and many times they are not effective, though it is not always understood why. A quick way to determine how effective the "group strap" is would be to measure the current flowing through the connection (perhaps focusing in on the specific frequency of noise that is attempted to be minimized). If there is no current flowing, then this indicates that the " group strap" is not working as intended. (the author has seen many examples of the need for "group straps" and whether the straps were installed or the straps were removed, the system had the same characteristics.) If this basic type of measurement would have been made, the cost, timing, and manufacturing impact of the installation of these connections could have perhaps been averted!

The concept of differential and common mode currents can be carried into higher frequencies that may be causing EMC issues in shipping systems. In this case, the current has been created by an external source of energy, such as nearby RF transmitter as shown in Fig. 7.8. Several key elements contribute to the magnitude of the possible problem:

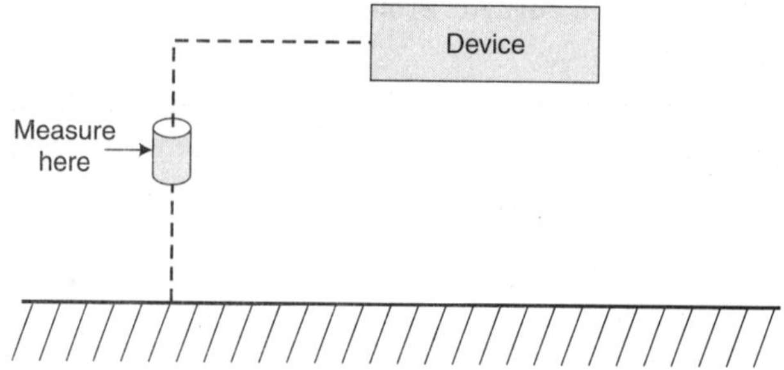

Fig. 7.7 Measuring differential mode current with a probe

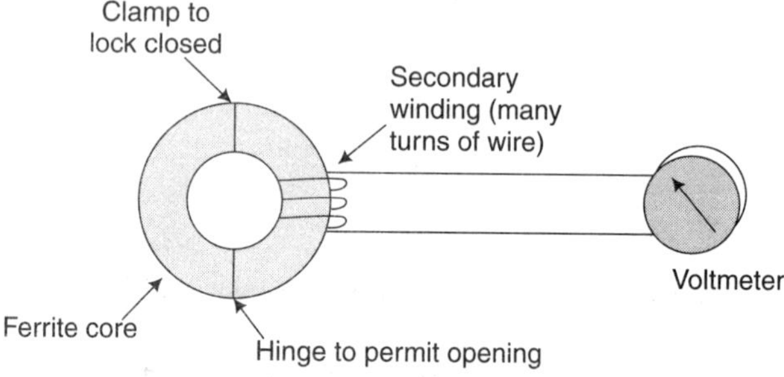

Fig. 7.8 Current probe electrical schematic

- The frequency of the external field
- The characteristics and the physical dimension of the conductors in the system that is being affected.
- The direction and magnitude of the external field.
- The source and the load impedance in the system that is being affected.

7.6.3 Cable Shielding

The major part of the coupling path for interference from the product to the environment and vice versa is through its connecting cables as shown in Fig. 7.9. These form efficient transducers with the outside world, via conduction at low frequencies and via radiation, particularly around their resonant frequencies (at which the cable length is multiple of a quarter wavelengths). These cables may actually intentionally carry high-frequency signals, such as data or video, but a more potent interference source is common-mode noise coupled onto the cable at the interface, and following in all its conductors or in its screen, which may not be directly related to the signal as shown in Fig. 7.10. A major part of EMC design is therefore concerned with the interfaces between the unit and its cables.

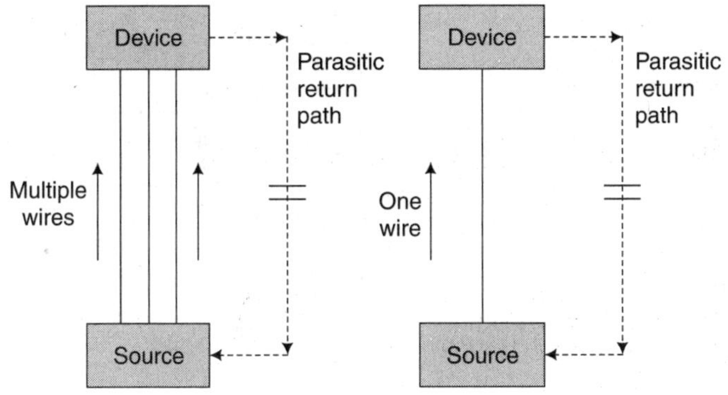

Fig. 7.9 Equivalent circuit

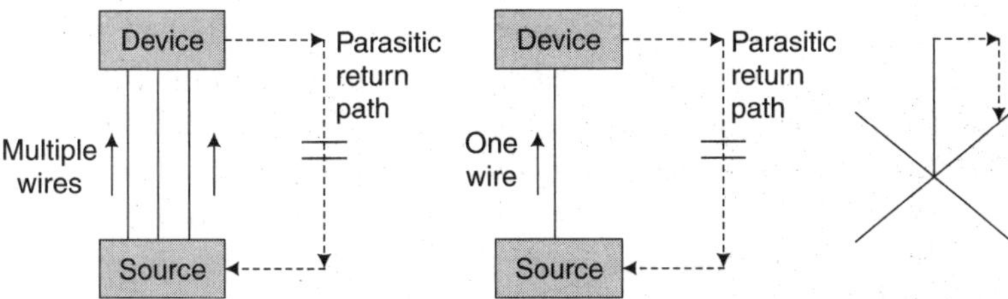

Fig. 7.10 Vertical antenna equivalent

One primary source of RE is unshielded or improperly shielded cables. There are four common types of shielding:

1. braid
2. flexible conduit
3. rigid conduit
4. spirally wound sheets of high permeability material

Of these four, braid is relatively lightweight and easiest to handle. It is important to note that the shielding effectiveness of a cable shield depends on the characteristics of the shield material and the manner in which the shields are terminated.

When terminating a shield, it is important that the termination provide a low impedance path for noise currents. Shield terminations fall into two categories: pigtail termination and 360 degree shield termination (sometimes referred to as RF back shell termination). The 360-degree shield termination provides a low impedance path and preservers shielding integrity of the enclosure or connector to which the shield is terminated. This type shield termination is much preferred. A pigtail termination is the least preferred method of shield termination because, at RF frequencies. The inductance of the pigtail becomes such that the shielding effectiveness of the cable shield is negated. If however, pigtail termination is unavoidable,

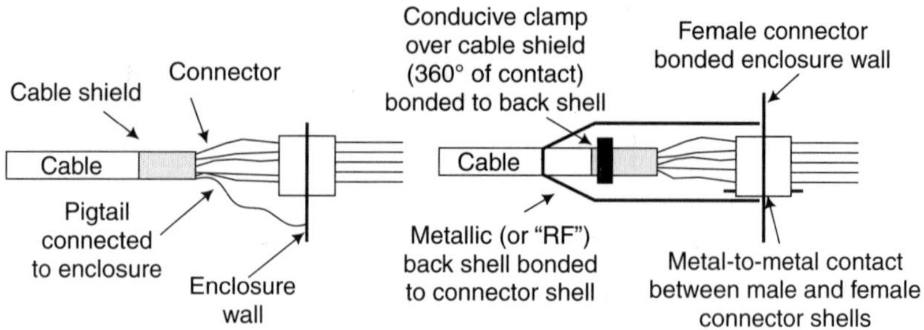

Fig. 7.11 Pigtail RF back shell terminations

keep the pigtail as short as possible. Figure 7.11 shows examples of pigtail termination and RF back shell termination.

Figure 7.12 shows the preferred methods of shield termination in descending order of preference. Specific requirements on cable shielding and shield termination are found in NSSA Handbook NHB 5300.4(3G).3-24. As a general rule, the cable shield should be grounded at both ends. Also, cable shields should never intentionally carry current. The exception to this rule is coax cable, in which the outer shield serves as the return conductor. Coaxial should be used only for signals where the lowest signal component is above approximately 100 kHz.

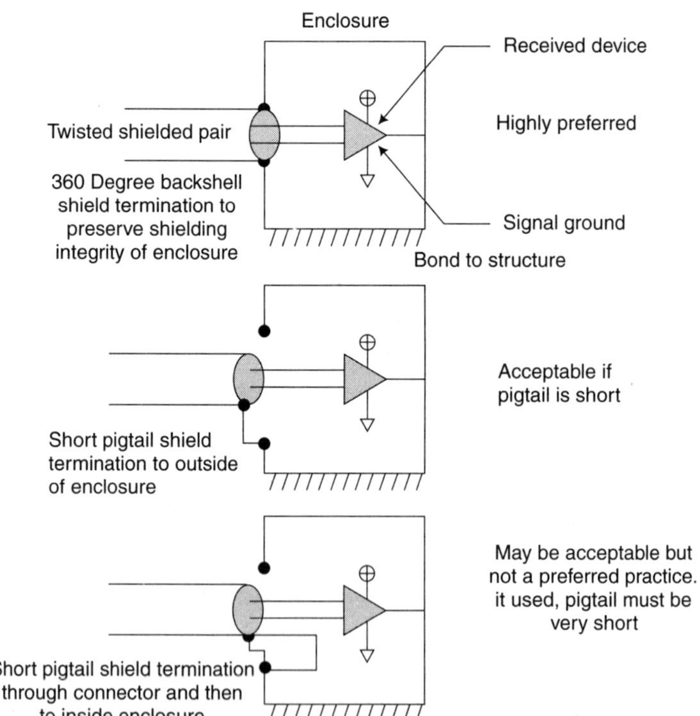

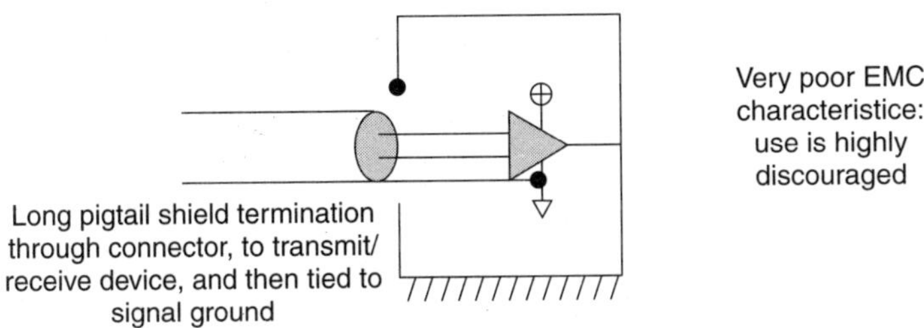

Fig. 7.12 Wire terminations

In some applications, double shielding of cables is required to prevent unwanted electromagnetic energy from entering the circuit. Figure 7.13 shows example schematics of how to ground double shielded cables. For some low-frequency, high-load-impedance circuits, grounding the shield at both ends causes low-frequency noise currents on the shield to couple low-frequency noise currents on the shield to couple into the circuit. Figure 7.14 is an example of a possible solution for this problem.

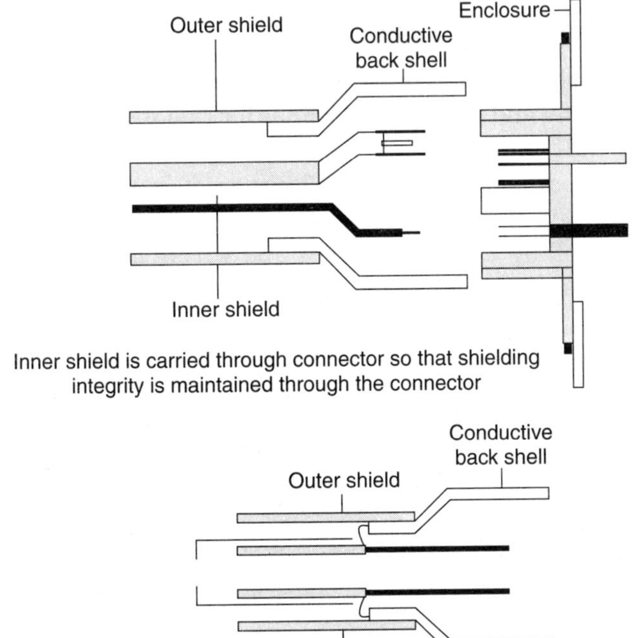

Fig. 7.13 Terminating double shielded cables

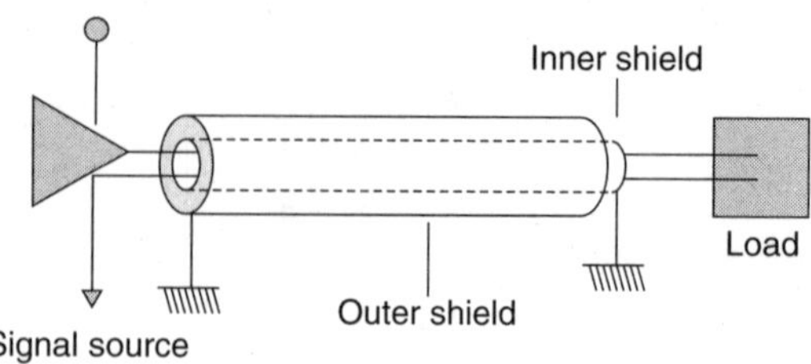

Outer and inner shields are grounded at one end and ungrounded at the other end. The two shields are isolated from each other at DC. At high frequencies, the capacitance between the inner and outer shield is such that the shield acts as if it were electrically grounded at both ends. This is effective against high frequency radiated fields

Fig. 7.14 Shielding low frequency, high impedance circuits

Outer and inner shields are grounded at one end and ungrounded at the other end. The two shields are isolated from each other at DC. At high frequencies, the capacitance between the inner and outer shield is such that the shield acts as if it were electrically grounded at both ends. This is effective against high frequency radiated fields.

7.6.4 Cable and Wiring Classes

Use of power and signal cables is prevalent on vessels. These cables may act as both transmitting and receiving antennas for radiated EMI and conduits for conducted EMI. Because cables are usually routed to accommodate practical routing paths and equipment location, it is almost impossible to predict and quantify the EMI environment associated with these cables. One way of controlling EMI from cables and wiring is to separate cables and wiring into similar classes of voltage, frequency, and susceptibility levels.

Much guidance in this area can be obtained from practices developed for the NASA space shuttle, international Space Station Alpha, and U.S. military. These specifications have requirements or guidelines for wiring classification and separation. For example, the U.S. Air Force Systems Command Design Handbook 1-4 suggest the classification of wiring based on type of electrical power (ac or dc) and frequency susceptibility. Also, as a design goal, Design Handbook 1-4 suggests a minimum separation of 2 in (51 mm) between different wire classifications to prevent cable-to-cable coupling. NASA specifications for Space lab payloads and space station program (MSFC-SPEC-521B, Electromagnetic Compatibility Requirements on Payload Equipment and Subsystems) and (SSP 30242, Space Station Cable/Wire Design and Control Requirements for Electromagnetic Compatibility) contain requirements for cable classifications and separation. For programs in which such requirements are not supplied,

Table 7.3 is a guide. The cables and bundles of different classifications should be separated by a minimum of 2 inches.

Table 7.3 Suggested cable classifications

Signal Type; Rise, Fall Time (tr, tf)	*Voltage or Sensitivity Level*	*Wire Type*	*Circuit Class*
Power (ac,dc)	>6 V	Twisted	Class I
Analog Signals tr,tf > 10 ms	>6 V	Twisted Shielded	Class II
Analog Signals tr,tf > 10 ms	≤ 100 mV	Twisted Doubled Shielded	Class III
Analog Signals tr,tf < 10 ms	≤ 100 mV	Twisted Doubled Shielded	Class IV
Analog Signals f > 100 kHz	A 11	Twisted Shielded, Coax	Class IV

Twisting the wire minimizes the loop area of the wire. This minimizes the amount of inductive noise coupling between the circuit and surrounding cabling. The number of twists per metre of cabling is limited by cable size; however, the greater the number of twists per metre, the smaller the loop area of the wire. Twisting will also increase the parasitic capacitance since the product of LC is a constant.

7.6.5 Vessel Generated Radiated Emissions

Shipping cabling systems can be sources of radiated emissions that may affect on-board electronics, or interfere with electronic systems on adjacent vessels or devices along the roadside, such as televisions, radios, etc. Active electronic devices generate radiated emissions during their normal operation because of switching operations, logic gates, PWM (pulse-width-modulated) control signals, etc. Even the non-solid state components, such as mechanical switches, horns, relays, other inductive devices, and spark plugs can also radiate. Many of these devices have been used since the early days of the shipping industry, and still require attention to ensure that measures are taken to prevent problems.

The shipping industry has classified emissions from electrical and electronic systems in two ways. They are classified BB (broadband) or NB (narrowband). Several methods can be used to review the material that is referenced at the end of this chapter. These classifications may vary as used in different industries. To understand the clarification one must understand the clarification used in the auto industry.

The auto industry for long has a straightforward approach to classification of the emissions (also called noise), which is as follows:

If the emission is created by an "arc" or "spark", it is classified as broadband (BB) noise. This means that all other devices are classified as "narrowband" (NB) emissions.

While this may not be the most rigorous method of classification, it is practical and has worked reasonably well for many years.

BB noise is emissions that occupy a wider frequency spectrum than the bandwidth of the receiver in use. This means that it is not possible to tune out of the emissions and tune to a frequency without emissions. However, with NB emissions, there may be frequencies that are occupied by the emissions sources and other frequencies that are not occupied by the noise. By looking at Fig. 7.15 we can understand how these terms relate to the actual physical process.

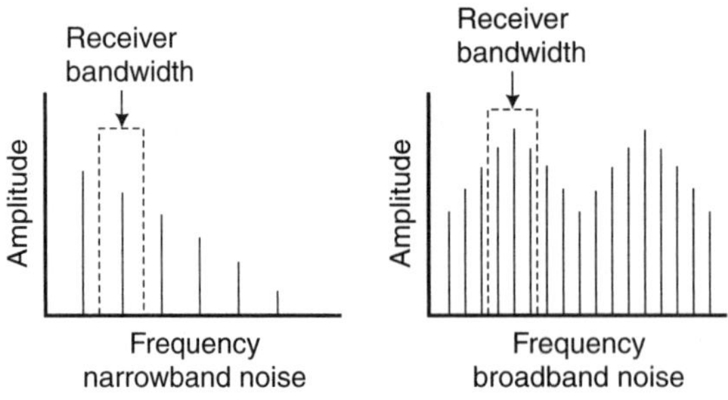

Fig. 7.15 Narrowband noise and broadband noise

Imagine that a sliding receiver "window" defines what is heard in the receiver. By moving this window across the BB noise in Fig. 7.15, you can see that since the emissions change as a function of time, there are always a significant amount of noise emissions in the receiver's bandwidth. Now, if the receiver's window is moved across the NB noise spectrum, there may be a setting where no noise falls within the receiver's pass band.

7.7 BROADBAND NOISE

7.7.1 Motor Noise

It is easy to observe the effects of BB noise from an electric motor. There are examples in every day applications, such as vacuum cleaners or mixers being operated in a home. It is not even necessary to have high power for the noise generation. Many "cordless" battery operated devices can emit enough energy to generate BB noise. There are many types/applications of motors in many environments, and of course, shipping systems also use many motors. These are used to adjust valve positions, open and close apertures, adjust pipe flows, in HVAC operation, and in power train control functions.

Another issue with motor noise that is that in addition to the energy being radiated from the motor itself, the conducted noise along the power feed from the battery or the emergency supply to the motor can contribute to radiated emissions. This is an issue in shipping systems because of the wiring that connects all loads to a common battery supply. The harness can look like a low impedance path for the noise energy.

7.7.2 Ignition Noise

An internal combustion spark ignited (SI) engine ignition system is a typical source of noise. Similar systems have been used since the beginning of the auto industry, and though some aspects of the technology have changed, the basic elements remain unchanged. Even if a small engine of the type is used on deck for a specific purpose it can create enough noise to disturb complex electronic systems.

7.7.3 SCR Noise

Another example of BB noise generation is from silicon controlled rectifier (SCR) controlled devices used widely in power control systems on board ships. Using a SCR to control the voltage to an AC operated device means that the sine wave is truncated at some point in its waveform. By doing this, the waveform starts to look like a portion of a square wave, with harmonic content. This can be heard by listening to an AM radio while an incandescent lamp dimmer is used to control a lighting circuit in the same room. Some of the newer light dimmers incorporate coils to "choke off" noise current from the accommodation wiring.

To review–the noise that is emitted from motors is BB. It appears on a spectrum display which will show many noise pulses within the instrument resolution bandwidth. These noise pulses will extend over a wide range of frequencies. Although most of the noise is generated by the motor brushes, even the "brush-less" motors can emit noise because of the switching of the current. The impact of electric motor noise on auto entertainment systems depends on the type of radio signal modulation. On AM radios, the noise will be heard as a "pop" or "click". In FM radios, the noise may not sound like noise, but may reduce the radio's sensitivity. The noise reduces the input SNR (signal-to-noise ratio). The reduced SNR will reduce the AM radio's output SNR. The FM radio may simply "quiet" at a higher input signal level, reducing the range over which a particular station may be received.

One of the characteristics of BB noise is that it is not possible to "tune out of" the noise. This can be seen with a simple demonstration using an AM radio and a source of BB emissions. These emissions can be from any type of high voltage discharge device (with a continuous discharge) such as a neon sign, a fluorescent light, or a motor. Tune into a station on the AM band, turn on the device, and note the interference. Tune the radio across the band and note how the noise still occurs at any frequency. This demonstration shows that there is a continuous band of energy that is being emitted across all frequencies.

7.7.4 Overview of BB Noise Sources

In summary, let's look at the characteristics and sources of BB noise. There are many systems and components in shipping that generate BB noise. Some of these components and technology date from the early days of the industry, and others are recent technology. The early components include spark plugs, ignition coils, switches, and motors. These can have EMC (mainly EMI) issues associated with them.

In order to address emissions from SI ignition systems, it may be easier to work at the component level, rather than at the vessel level. For example, the radiated emissions level from spark plug RFI was reduced in the 1950s to achieve compatibility with radio receivers, and these techniques are still relevant today.

7.8 NARROWBAND NOISE

Let's review the characteristics of NB noise at this point. Recall from earlier discussions that we classify noise as NB if its bandwidth is less than the affected receiver's RBW. This means that there may be only one or a few frequencies emitted from the device that will interfere with a radio receiver. The nature of NB emissions is that the frequencies at which the noise is observed and the amplitude of the noise tend to be stable over time.

7.8.1 Microprocessors and Narrowband Noise

The following is an example of how the emissions from a typical microprocessor source could occur, as result of harmonics of the fundamental frequency that are emitted from the microprocessor's operation. If we look closely at this, we will see that there is a very repeatable pattern that occurs with the emissions. This very repeatable pattern occurs at multiples of the clock frequency. What is significant about this type of emissions is that the emissions will stay constant with respect to amplitude overtime, unlike broadband emissions where both the amplitude and the frequency will vary. This is referred to as a "comb like" appearance, because the emission pattern looks like an inverted comb.

One of the major sources of NB emissions in today's shipping is the microprocessor. In order to operate a digital microprocessor; we require clock and logic signals of sufficient amplitude. Secondly, microprocessors are being added in greater numbers to more vessel modules. Lastly, designers of digital circuits prefer short pulse rise and fall times to minimize timing uncertainly and to reduce device heat dissipation. If we look at the spectral content emitted from microprocessor operation, it may resemble the emissions at the harmonics of the clock fundamental frequency. Many of you may recall you learned that only "odd harmonics" would exist in a square wave. This is true only for exactly 50% duty cycle signals, which seldom occur in the real world!

Microprocessors are not the only source of the NB emission. Many other devices, such as power transistors, PWM (pulse width modulated) speed controls and switching transistors can generate NB emissions. These are widely used in winch gear controls, windlass controls, boiler controls, watertight door control, specialized towing and hauling gear control.

7.8.2 Generation of Narrowband Interference

NB interference can be demonstrated very easily. Using a portable FM broadcast receiver, tune to a situation in the FM band. Create an emission at that same frequency using an NB source (with a stable oscillator, or a RF signal generator). If the signal strength from the received station is higher then the received station will be blocked out by the emission from the test source. Note that, while this particular station would be blocked out, the radio would appear to function normally on all the other stations. This can be very confusing from a diagnostic process standpoint, as it will look like the receiver and the system are operating normally except at that one particular frequency. This may also be accomplished by using a second FM radio tuned 10.7 MHz below the receive frequency of the first radio.

7.9 NARROW BAND RADIATED EMISSIONS CASE STUDY

Just how much concern is radiated emissions from microprocessor based devices today?

Let's look at an actual case of NB radiated emissions on a vessel. This shows what can happen when unanticipated radiated emissions fall on a frequency that is being utilized for vessel operation.

The problem occurred on a coastal vessel. The two-way radio used to communicate with the shore used frequencies in the VHF band. The radios could not be utilized because radiated emissions from the vessel were on the communication channel frequency. To analyze the path that the emissions coupled to the radio, a series of tests was performed. By disconnecting the vessel power from the radio and powering the radio from battery, it was determined that the noise was not being conducted through the power leads or other wiring. By disconnecting the antenna cable from the radio, it was determined that the noise was being radiated from the engine control system to the antenna for the radio.(this process can be very helpful in diagnosing vessel radio problems).

Several solutions were tried and the one that was changing the frequency of the microprocessor clock reference oscillator crystal. Since the crystal fundamental frequency is multiplied many times to generate the VHF band harmonic, a very small change in the fundamental frequency can yield a significant change at the frequency of concern.

7.9.1 Impact Narrowband Noise

In summary, NB noise sources may affect only specific frequencies, which can result in receivers appearing to function normally. However NB noise may often be addressed in the component design process by obtaining data early in the design or development process. By selecting a "Quiet" microprocessor and designing the circuitry to minimize NB emissions by increasing the impedance or providing "shunt path" for the emissions, radiated emissions problems may be minimized

Again, there are three characteristics of narrowband noise sources:

- NB noise may only affect specific frequencies.
- Receivers can appear function almost normally in the presence of NB noise.
- NB emissions can be addressed in component design process.

NB noise and radiated emissions have known to shutdown crankcase mist detectors, boiler flame control device, Inert Gas control device, aux engine and aux machinery control devices. In some cases there are recorded evidence of accidents occurring and in others they have been the secondary causes. This is an area still unaddressed in shipping circles, but better shielding techniques and use of more stringent emission norms are making it safer.

Annexure 1

Classification and Protection of Cables

The various conditions proposed to be adhered to is laid down in Table D1, D2, D3 & D4

Table A1 Minimum clearance and creepage distance

	Clearances mm				Creepage distances mm	
System	*0 ≤ 63A*		*0 ≥ 63A*		*J ≤ 63A*	*J ≥ 63A*
Voltage V	L-L	L-A	L-L	L-A		
Up to 60	2	3	3	5	3	4
61-250	3	5	5	6	4	8
251-380	4	6	6	6	6	10
381-500	6	8	8	10	10	12
501-660	6	8	8	10	12	14
661-750	10	14	10	14	14	20
751-1000	14	20	14	20	20	28

L-L Between two live parts and between a live part and an earthed part.

L-A Between a live part and a part which can become accidentally dangerous,

Table B1 Parameter class for the different locations on board

Parameter	*Class*	*Location*
Temperature	A	Machinery spares, control rooms, accommodation bridge
	B	Inside cabinets, desks etc. with temperature rise of 5°c or more installed m location A
	C	Pump room, holds, rooms, with no heating
	D	Open deck, masts and inside cabinets, desks etc. with a temperature rise of 5°C or more installed in location C.
Vibration	A	On bulkheads, beams deck, bridge
	B	On machinery such as internal combustion engines, compressors pumps, melding piping on such machinery
	C	Masts

Table C1

Class	*Location*	*Degree of Protection*
A	Control rooms, accommodation, bridge	IP 22
B	Machinery Space	IP 44
C	Open deck, masts, below floor plates in machinery space	IP 56
D	Submerged application	IP 68

Table D1 Cable classes

Class	*Classification*	*Function (examples)*
A	EMI generating Not EMI susceptible 24V-600V, DC 50-60Hz, 400Hz High power upto 10Kv	Power cables Control cables in components using mechanical contacts and relay coil.
B	Slightly EMI generating Slightly EMI susceptible 0.5 V – 50V. low frequency	Telephone cables Signal cables Synchro circuits (60 – 400Hz)
C	EMI generating EMI susceptible 0.1-5V 50W pulse 0.1-24V DC	Video signals Data transmission Analog measuring values after converter
D	Highly EMI susceptible 10m V-100mV 50 – 2000 W DC. AF. HF	Receiver Antennas, Hydrophone and microphone cables Analog measuring values
E	Highly EMI generating	Radio transmitter Solar Radar modulator Thyristor controls

Table D2 Minimum distances in meters between unscreened cables of different classes when paralleled over more than 2 meters

Cable Class	*A*	*B*	*C*	*D*	*E*
A	0.00	0.25	0.25	0.50	0.25
B	0.25	0.00	0.25	0.25	0.25
C	0.25	0.25	0.00	0.25	0.50
D	0.50	0.25	0.25	0.00	0.50
E	0.25	0.25	0.50	0.50	0.00

Table D3 Minimum distances in meters between unscreened cables of different classes when crossing at 90°c

Cable Class	*A*	*B*	*C*	*D*	*E*
A	0.00	0.00	0.15	0.30	0.15
B	0.00	0.00	0.15	0.15	0.15
C	0.15	0.15	0.00	0.00	0.30
D	0.30	0.15	0.00	0.00	0.30
E	0.15	0.15	0.30	0.30	0.00

Table D4 Minimum distances in meters and grounded cables of different cable between unscreened ones

Cable Class	*A+B*	*C+D*	*E*
A+B	0.00	0.10	0.25*
C+D	0.10	0.00	0.50*
E	0.25*	0.50*	0.00

* For steel tube or conduct having a wall thickness of at least 1.5 mm around class E cables. Delete the separation requirement.

Annexure 2

Coupling Paths on-Board Ships

Fig.1 Antenna Arrangements

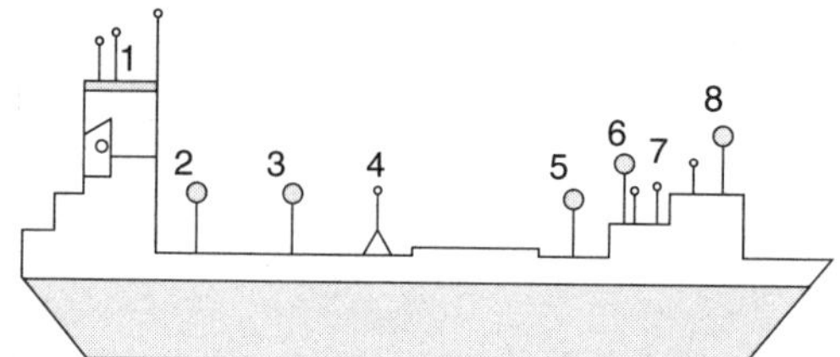

Fig. 2 Complexity of Antennas

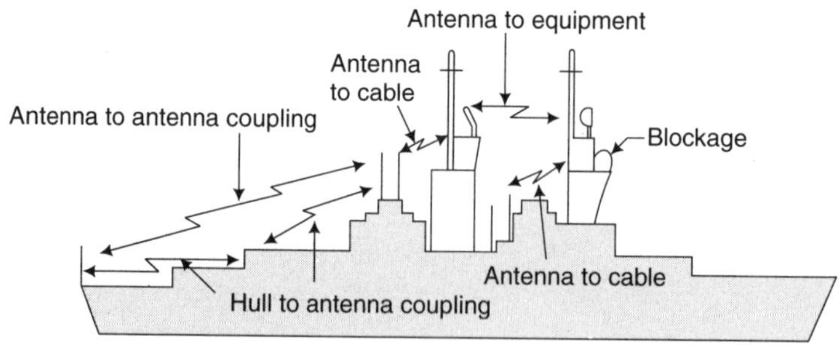

Fig. 3 Coupling Paths

Annexure 3

Communications Mesh on Bridge

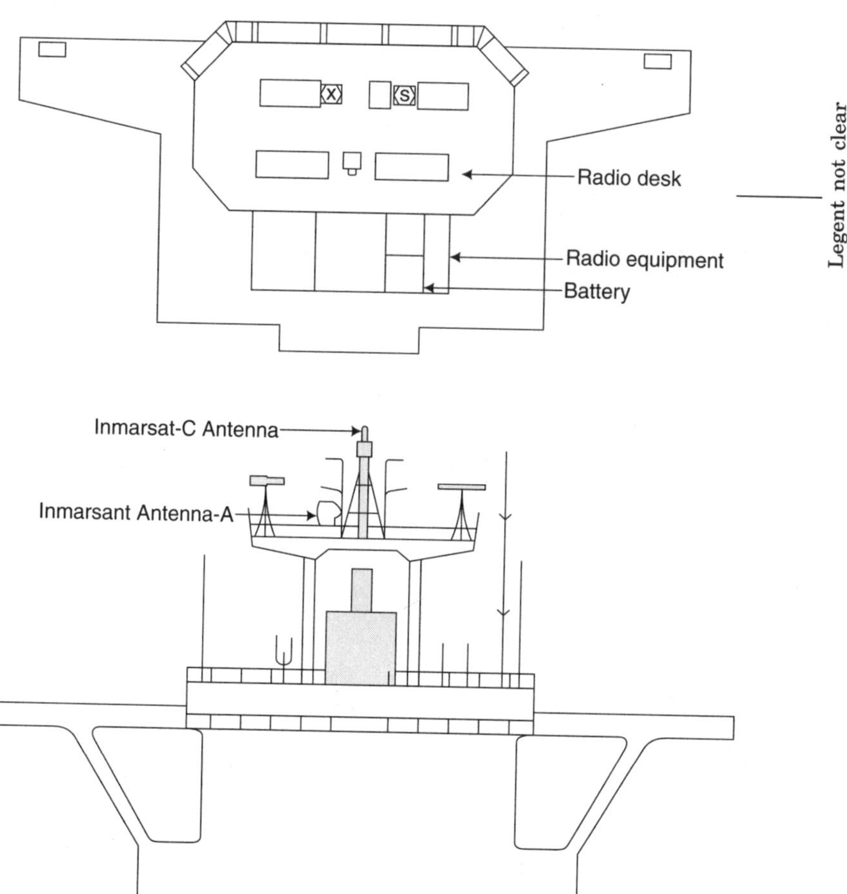

Legent not clear

Annexure 4

Flow Chart for Modeling

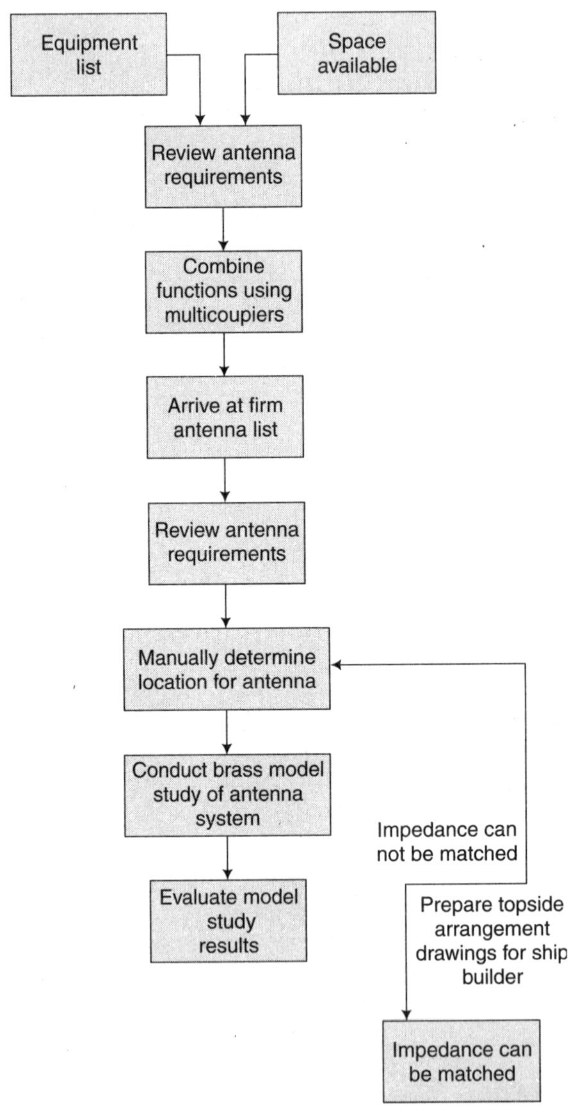

Annexure 5

EMC Check List

A. Suppressing noise source enclosure		
Item	*Y/N*	*Comments*
Enclosure noise source in a shielded enclosure		
Filter all leads leaving a noisy environment		
Limit pulse rise times		
Relay coils should be provided with some form of surge damping		
Twist noisy leads together		
Shield and twist noisy leads		
Ground both ends of shields used to suppress radiated interference (shield does not need to be insulated)		

B. Eliminating noise coupling		
Item	*Y/N*	*Comments*
Twist low-level signal leads		
Place low-level leads near chassis (especially if circuit impedance is high)		
Twist and shield signal leads (coaxial cable may be used at high frequencies)		
Shielded cables used to protect low-level signal leads should be grounded at one end only.		Coaxial cable may be used at high frequencies with shield grounded at both ends
When low-level signal leads and noisy leads are in the same connector, separate them and place the ground leads between them		
Carry shield on signal leads through connector on separate pin		
Avoid common ground leads between high and low level equipment		
Keep hardware grounds separate from circuit grounds		
Keep ground leads as short as possible		

B. Eliminating noise coupling		
Item	*Y/N*	*Comments*
Use low-impedance power distribution lines		
Avoid ground loops		Consider using the following devices for breaking ground loops: Isolation or neutralizing transformer; Optical couplers; Differential amplifier; Guarded amplifier; Balanced circuits
Use steel cabinets of 1-2mm thickness		Avoid using plastic enclosures
Aluminium superstructures to be grounded by CUPAL – foils every 10m		
When coaxial cables are used the cable screen may be used as a return and ground		It may be advantageous to use a filter to prevent ground loops
Avoid coaxial cable with BNC connectors		Use preferably twinax or traix having an outer screen that does not carry signal
Avoid using RS 232 for data communication		RS232 is not balance i.e. conductors do not have the same impedance with respect to ground and to all other conductors. RS485 is better. TTY current loop 20mA is balanced, sender and receiver are separated (low capacity).
Proper connection and clean the surface		Use stainless steel bolts, zinc spray, lock washers, etc.
Copper straps for grounding should have a cross section of at least $25mm^2$		Use at least one strap per metre of the cabinet and preferably at both bottom and top
Use conductive coating in place of non conductive coatings for protection of metallic surfaces		
Separate noisy and quiet leads		
Ground circuits at one point only (except at high frequencies)		
Avoid questionable or accidental grounds		

B. Eliminating noise coupling		
Item	*Y/N*	*Comments*
For very sensitive application operate source and load balanced to ground		
Place sensitive equipment in shielded enclosures		
Filter or decouple any leads entering enclosures containing sensitive equipment		
Keep the length of sensitive leads as short as possible		
Keep the length leads extending beyond cable shielded as short as possible		

C. Reducing noise at receiver		
Item	*Y/N*	*Comments*
Use only necessary bandwidth		
Use frequently selective filters when applicable		
Provide proper power supply decoupling		
Bypass electrolytic capacitors with small high-frequency, capacitors		
Separate signal, noisy, and hardware grounds		
Use shielded enclosures		

Information Sheet for EMC Management

Information Sheet for EMC Management

Project: System: Reference: Drawing Date Sheet:	Page: Sign: Date:		
Item Reference No.:	Characteristic:		
Specifications	*Emitter*	*Coupling*	*Receiver*
Equipment, System Operational: What is the system intended to do? When is the system operative? Stand alone, or part of a larger system? With what equipment does the system interface directly and indirectly? Is the system required to operate continuously or intermittently? Are there critical sequences of operations involving this system? How will the system be maintained, operated, and supported? Installation details: Location (type of facility in which the equipment to be installed. E.g. machinery space, accommodation) What other equipment will be in the same installation? Cable specification type Class, Routing Single point or multi point grounding Protection against electrostatic discharge?			

Contd...

Contd...

Type of filter applied Input Output Power supply Requirements: Basic power requirements? Signal inputs, and their range of frequency and power? Signal outputs, and their range of frequency and power? Sensitivity requirement for receiving equipment? Requirement to emission of disturbance Requirement to total harmonic distortion (%) Requirement to electrostatic discharge ESD (kV) Consequence: Consequence of failure Consequence class Conclusions: Decoupling level (dB) Distance-dependent reduction EM environment platform be? Radiation (dBuV/m) Conduction (dBuA/m) Frequency (MHz) Anticipated attenuation by additional means (dB) Consequence class (0.6,10,20dB) Required susceptibility (dB) Specified susceptibility (dB) Estimated margin between noise and susceptibility (dB) Susceptibility to harmonic distortion (Yes / No) Susceptibility to ESD (Yes / No)			

References

1. **Clayton R. Paul** - Introduction to Electromagnetic Compatibility, John Wiley & Sons, Inc, USA, 1992.
2. **Terence Rybak** and **Mark Steffka** - Shipping Electromagnetic Compatibility (EMC), Kluwer Academic Pulishers, USA, 2004
3. **Dipak L. Sengupta** and **Valdis V. Liepa** - Applied Electromagnetic and Electromagnetic Compatibility, John Wiley & Sons, Inc., USA, 2006
4. The ARRL RFI Book - Practical Cures for Radio Frequency Interference, The American Radio Relay League, USA, 1998 – 1999.
5. **C.A. Balanis** - Antenna Theory Analysis and Design, John Wiley & Sons, Inc., USA 1982.
6. **P.M. Rao, G.S.N. Raju** "Some Studies on arrays for side lobe reduction required for EMI control", Proc. INCEMIC, pp 314 – 320. Dec. 1995.
7. Mark I. Monstrose and Edward M. Nakauchi - Testing for EMC Compliance Approaches and Techniques, John Wiley & Sons, Inc, USA, 2004.
8. **V.P. Kodali** - Engineering Electromagnetic Compatibility, IEEE Press, New York, 1988.
9. **A.A. Smith** - Jr. Radio Frequency Principles and Applications, IEEE Press, The Institute of Electrical and Electronics Engineers, New York, 1998.
10. **D.A. Morgan** - A Handbook for EMC Testing and Measurement Series 8, Peter Peregrinus, London, 1994.
11. **P.Misra** - Shielding effectiveness of some material for EMC design of equipment in ships - Maritime Symposium IME, Chennai, 1996.
12. **P. Misra** - EMR Effect and EMC Considerations For Improving Safety Index Of Fishing Vessels. – International Conference On EMI 16 to 17 December – 1999, New Delhi.
13. **P. Misra** - EMR effects and EMC considerations for improving safety index of fishing vessels. – MER, Mumbai '1999.
14. **P. Misra** - EMR Effect and EMC Considerations for Improving Safety Index Of Fishing Vessels. – International Conference on EMI 16 to 17 December – 1999, New Delhi.
15. **P. Misra** - Estimation of Shielding Effectiveness of Some Typical Materials – Seminar of EMI Problems and Design for EMC 13-14 March '97, Calcutta.

16. **Cleary, S.F.** (1973) Uncertainties in the Evaluation of the Biological effects of microwave and radiofrequency radiation.
17. **Engen, G.F.** (1973) Theory of VAF and Microwave Measurements Using The PowerEquation Concept. Washington, Dc, National Bureau Of Standards. (NBS Tech. Note 637, April 1973).
18. **Gordon, Z.V.** (1970) Occupational Health Aspects Of Radiofrequency Electromagnetic Radiation. In: Ergonomics and Physical Environmental Factors. Geneva, International Labour Office, Pp. 159-174 (Occupational Safety And Health Series No. 21).
19. **Hankin, N.N.** (1974) An Evaluation Of Selected Satellite Communication Systems As Sources Of Environmental Microwave Radiation, Silver Spring, MD, US Environmental Protection Agency, 56pp. (Report EPA – 520 / 2-74-008).
20. **Hunt, E.L., King, N.W., & Phillips, R.D.** (1974) Behavioral Effects of pulsed microwave radiation.
21. **Johnson, C.C. & Shore, M.L.,** Ed. (1977) Biological Effects Of Electromagnetic Waves, Symposium Proceedings, Boulder, October 1975, Rockville, US Dept of Health, Education, and Welfare, 693 Pp. (FDA, BRH, Publication HEW-FDA 77.8010).
22. **H. Bassen** and **R. Peterson**, "Complete Measurement Of Electromagnetic Fields With Electro-Optical Crystals," In Biological Effects Of Electromagnetic Waves, C.C. Johnson, Ed., Vol. Ii, P.310 (1970).
23. **F.W. Shaw, L. R. Shuté**, and **S.T.Li**, "Shipboard RF Environment and Requirements for Monitoring Emcon Status on the USS Germantown (LSD - 42."Nosc in 1673(1991).
24. C.H.Coz, L.A. Bernotas, G.E. Betts, A. I. Grayzel, D.R. O'brien, J.J. **Scozzafava**, and **Ac. Yee**, "An Externally-Modulated Fiber-Optic Link Test Bed For Addressing Application Issues and Options," PSAA-91 Techdig. 1 (1992)
25. **Paul R. Clayton**; Introduction To Electromagnetic Compatibility: John Wiley & Son. Inc. 1992
26. IEEE Std. 519-1992 Recommended Practice and Requirements for Harmonic Control on Electrical Power Systems.
27. Mil-Hdbk-237a, Electromagnetic Compatibility Management Guide for Platform Systems and Equipment, (1981).

Index

R

S

T

U